Revolutionary Mathematics

Revolutionary Mathematics

Artificial Intelligence, Statistics and the Logic of Capitalism

Justin Joque

VERSO
London • New York

First published by Verso 2022
© Justin Joque 2022

All rights reserved

The moral rights of the author have been asserted

1 3 5 7 9 10 8 6 4 2

Verso
UK: 6 Meard Street, London W1F 0EG
US: 20 Jay Street, Suite 1010, Brooklyn, NY 11201
versobooks.com

Verso is the imprint of New Left Books

ISBN-13: 978-1-78873-400-4
ISBN-13: 978-1-78873-402-8 (UK EBK)
ISBN-13: 978-1-78873-401-1 (US EBK)

British Library Cataloguing in Publication Data
A catalogue record for this book is available from the British Library

Library of Congress Cataloging-in-Publication Data
A catalog record for this book is available from the Library of Congress

Typeset in Minion by Hewer Text UK Ltd, Edinburgh
Printed and bound by CPI Group (UK) Ltd, Croydon CR0 4YY

Contents

Acknowledgments vii

Introduction 1

Part I: Ghosts of Departed Quantities

1. Automating Knowledge 37
2. Can Computers Do Math? 59
3. Algorithms of Objectification 77

Part II: The Promise of Frequentist Knowledge

4. Do Dead Fish Believe in God? 101
5. Induction, Behavior and the Fractured Edifice of Frequentism 115

Part III: Bayesian Dreams

6. Bayesian Statistics and the Problem with Frequentism 143
7. Bayesian Metaphysics and the Foundation of Knowledge 158

8. Automated Abstractions and Alienation 177

Conclusion: Toward a Revolutionary Mathematics 217
Index 225

Acknowledgments

This book was written in what seemed like a fever dream, or more accurately a series of overlapping dreams. Such a series of dreams may appear anathema to any attempt to account for the mechanistic computation of statistics and algorithms, but they illustrate precisely what this text attempts to show: these calculations are passionate, and no less mysterious than the most revelatory of religions. They are social and mystical sciences that guide us in turning collections of data into knowledge of the world. So, if there is to be any hope of breaking the current alliance between capitalist objectivity and statistics, we must not shy away from what Marx describes as the "mist-enveloped regions of the religious world"— or, in our case, the mist-enveloped regions of statistics and machine learning.

A whole host of individuals have floated through this text and its dreams at various points; everything that is commendable or interesting or inspiring within it owes every bit of thanks to them. Christoph Becker, Maria Angela Ferrario, David Phillips, Alessandro Delfanti and Costis Dallas read an early draft of the book and spent a beautiful spring day in Toronto indoors offering immensely helpful feedback, for which I am deeply grateful. Laura

Portwood-Stacer provided significant developmental suggestions that helped this book along its way. The intellectual generosity of Petero Kalulé and our many far-ranging conversations were incredibly helpful as I worked on this project. Robyn Anspach, my wife, who I met by way of an algorithm, also spent considerable time with this text and helped guide it into its final form. Cengiz Salman worked tirelessly throughout this project assisting in a variety of ways, from research to editing to helping draft sections. He was especially critical in thinking through and writing the final chapter, which thus bears his name as well.

Finally, this book would not have been possible without John Cheney-Lippold. Originally, we intended to cowrite this book, and we spent many long afternoons at bars and cafés around Ann Arbor discussing, thinking, writing and revising. We worked for a couple of years like this, but toward the end, John became involved in a political situation that required all of his energy and time. He graciously encouraged me to finish the book on my own. This book owes its existence to him and his friendship. His work and name are here in absentia, even if it is not signed.

Introduction

> Men make their own history, but they do not make it as they please; they do not make it under self-selected circumstances, but under circumstances existing already, given and transmitted from the past. The tradition of all dead generations weighs like a nightmare on the brains of the living.
>
> —Karl Marx, *The Eighteenth Brumaire of Louis Bonaparte*

Computers are amazingly efficient at producing knowledge. Through surveillance and monitoring architectures that record the most miniscule of events, computers are able to track the stars, billions of people, transportation and logistics networks, trillions of dollars' worth of financial transactions, and exabytes of written words, audio and video. This data does not sit there inertly. Using this data and a combination of machine learning, statistics and raw processing power, corporations, governments and researchers are able to model, predict and infer everything from the probability of major policy shifts to our individual habits and desires. As a result, our interactions are becoming personalized, from shopping to cancer treatments, where treatment now can depend on the genetic makeup of a patient's cancer cells.

Not only is computation changing how we understand physical, social and economic systems; it is also reshaping our understanding of science and knowledge itself. Science increasingly involves the collection of huge amounts of data—telescope imagery of the night sky, the human genome, millions of people's health data—that is used to algorithmically seek out clusters and patterns.

It is clear that machine learning, algorithms and statistics shape our situation today.[1] But, as these technologies have developed, a key incongruity has arisen. While corporations and governments are able to predict and categorize individual life and actions into more and more fine-grained subgroups—a power that should formally allow for more efficient and just distribution of goods and opportunities—technology has instead been turned against humanity's very survival. As it has been sold to us, the cyberutopian impulse that we need more of this cybernetic system, that more data and better algorithms will resolve its contradictions, is clearly in the wrong.

From the cold caverns of corporate cloud storage facilities to the high-security citadels of the National Security Agency and its UK counterpart, the Government Communications Headquarters, the use of statistics transforms massive stores of digitized data into actionable information for human and computer consumption. This information is used to trade stocks, adjust prisoners' sentences, grant or deny credit and infer scientific facts. Computers today can do and understand things that only ten years ago seemed like science fiction. Even without the realization of full "artificial

---

[1] For Althusser, the process of perceiving the current situation is a process by which one is able to "act on History from within the sole history present," working on "what is specific in the contradiction and in the dialectic . . . not to demonstrate or explain the 'inevitable' revolutions post festum, but to 'make' them in our unique present, or, as Marx profoundly formulated it, to make the dialectic into a revolutionary method, rather than the theory of the fait accompli." Louis Althusser, "On the Materialist Dialectic: On the Unevenness of Origins," in *For Marx*, trans. Ben Brewster (London and New York: Verso, 2005), 180.

intelligence," machines are nonetheless astoundingly capacious. But at the same time these technologies allow those who own and operate them more and more freedom and insight, the political and social order is increasingly subject to the constraints of a closely managed global neoliberalism—even if it is an inefficient one, full of explosive contradictions.[2] Everywhere—according to various national, gendered, classed and raced positionalities—populations are surveilled and tracked, their perceptions and politics managed through highly individualized advertising and information selection. The only force that appears capable of breaking with this order is a death drive of extremism, from Donald Trump to Brexit to the Islamic State. And in the end, even this force seems to play back into the governing logic of managerial capitalism, as every crisis, whether flooding, terrorism or the rise of right-wing nationalism, generates the same calls for the retrenchment of an increasingly privatized capitalism.[3]

The German Marxist Alfred Sohn-Rethel hoped that human society would undergo "the transition from the uncontrolled to the fully conscious development of mankind."[4] While we are witnessing the development of technologies that, according to their proponents, should allow such a conscious development, we are simultaneously experiencing the further destitution of this transition's political possibility. Perhaps such a transition has always been impossible, but technology appears to be accentuating this uncontrolled and dangerous development. One of the clearest examples of this is the logic of austerity politics over the last few decades and neoliberalism's attendant moves toward privatization—a logic of saving and economic conservatism that has served only the

---

2 Wendy Brown, "American Nightmare: Neoliberalism, Neoconservatism, and De-democratization," *Political Theory* 34, No. 6 (2006): 690–714.

3 Naomi Klein, *The Shock Doctrine: The Rise of Disaster Capitalism* (London: Macmillan, 2007).

4 Alfred Sohn-Rethel, *Intellectual and Manual Labour: A Critique of Epistemology* (Atlantic Highlands, NJ: Humanities Press, 1978), 2.

profiteering of capitalist institutions.[5] Since 2008, algorithmic trading—that is, algorithms that trade stocks among themselves—has accounted for at least half the average daily volume in US stock markets.[6] Such automation has done nothing to make capitalism more rational; instead it has created a technological arms race that only accentuates the power of ratings agencies and investors to make demands on states to slash budget outlays, under threat of severe economic retribution. While rationality and truth have long been in crisis, their condition is especially dire today, with the weight of these crises recognized even in the halls of power.

At the same time that these utopian and dystopian visions of a world watched over by algorithms saturate much writing on technology, statistics and machine learning are also in deep crisis: academic knowledge production is beset by unreplicable studies, the carbon impacts of big data infrastructures add to threats of environmental devastation, and artificial intelligence is possibly poised for a new winter of abandonment.[7] Despite the meteoric growth of machine learning and data analytics—and the trillions of dollars generated in its wake—algorithmic knowledge production seems unable to keep pace with the motives and drives of capitalist production.

We need only look at the US Environmental Protection Agency's announcement, in September 2015, of its discovery that Volkswagen was selling diesel cars containing a "defeat device." This device, installed on 11 million different models between 2009 and 2016, was a piece of software that detected when a car was undergoing a

---

5 See, for example, United Nations Conference on Trade and Development, *Trade and Development Report 2017: Beyond Austerity; Towards a Global New Deal*, 2017.

6 Robin Wigglesworth, "The Quickening Evolution of Trading-In Charts," *Financial Times*, April 11, 2017.

7 Andrew Gelman, "The Problems with P-values Are Not Just with P-values," *American Statistician* 70 (2016); Filip Pieniwski, "The AI Winter Is Well on Its Way," *Venture Beat*, June 4, 2018; John Harris, "Our Phones and Gadgets Are Now Endangering the Planet," *Guardian*, July 17, 2018.

state-mandated laboratory emissions environmental test.[8] If the car believed it was being tested, Volkswagen's software would alter the engine's performance to comply with environmental regulations. This allowed Volkswagen's cars to be fast and powerful when driven under normal conditions—producing nitrogen oxide levels upward of forty times the legal limit—but fuel-efficient and environmentally responsible when undergoing a test. In the wake of the scandal, six Volkswagen executives were criminally charged.

Such a defeat device was possible only through the use of software; a physical device likely would have been detected much earlier. Hidden within each car's central processing unit, Volkswagen's software detected "the position of the steering wheel, vehicle speed, the duration of the engine's operation, and barometric pressure" in order to decide whether to limit emissions or not.[9] This type of deception is a direct threat to the supposed economic rationality of capitalism and the ability of states to manage and maintain markets. The US district judge in charge of the case, Sean Cox, declared at the time of sentencing that this crime "attacks and destroys the very foundation of our economic system: That is trust."[10] In short, by attacking the solidity of our knowledge of the world, these defeat devices threaten to undermine the supposed metaphysical stability of markets.

Two years later, the *New York Times* reported that ridesharing app Uber had created a program, code name "Greyball," that from 2014 to 2017 detected users who were government regulators and refused them rides.[11] Like Volkswagen, Uber's program combined data—such as location (near or far from a government building), social

---

8 Jack Ewing, "Volkswagen Says 11 Million Cars Worldwide Are Affected in Diesel Deception," *New York Times*, September 22, 2015.

9 James Grimmelmann, "The VW Scandal Is Just the Beginning," *Mother Jones*, September 24, 2015.

10 Sarah Ruiz-Grossman, "Volkswagen Executive Gets Max Sentence of 7 Years for Role in Emissions Scandal," *Huffington Post*, December 6, 2017.

11 Mike Isaac, "How Uber Deceives the Authorities Worldwide," *New York Times*, March 3, 2017.

media accounts, credit card information, cell phone numbers or device IDs of known law enforcement agents, and even the frequency with which users opened and closed the Uber app—to assess if a user was a potential regulator. If that user was deemed a "regulator," the app would "hide the standard city app view for individual riders, enabling Uber to show that same rider a different version."[12] While normal users could see drivers in the app that could potentially pick them, regulators were blocked from seeing these same drivers—and hence unable to regulate them or the company.

For a company like Uber, whose business strategy is ostensibly to operate in a legal gray zone, undersell transportation services and eventually monopolize a given area's taxi sector, Greyball was especially useful in places like Austin, Texas, and Portland, Oregon, where local laws made Uber illegal. By using Greyball as a tool to avoid regulation, the company minimized its own risk. Uber used the program to shield itself not only from American authorities, but also from the Australian, Chinese, and South Korean governments.

Programs like Volkswagen's and Uber's require us to radically reconsider the fundamental presumptions of algorithmic logics and the so-called big data revolution. Both proponents and opponents of massive data collection traditionally present big data as a way to understand the world: big data advocates see data and algorithms as a means to a better, more rational society, while its critics fear that the insights furnished by these systems will be used to control society, eroding freedoms. But when these technologies can be used to present the appearance that one is following rules while wholly disregarding them, the presumed veracity of algorithmic knowledge production falls apart. The ability to understand and shape the world of appearances soon becomes a key area of competition, where the state can only regulate those companies whose engineers and programmers fall far enough behind to get caught.

---

12 Joe Sullivan, "An Update on 'Greyballing,'" *Uber Newsroom*, March 9, 2017.

Even beyond direct attempts at duplicitousness, the rise of algorithmic capitalism has created an economy where statistics and algorithms are more efficient at creating new realities (e.g., the virtual world of derivatives trading, the platforms of social media and contract labor, and the private social world of filter bubbles) than they are at representing a world.[13] These methods of "re-simulation" frame economic decision making in a distinct light, one where the earnestness of capitalist idealism is usurped by threats from all sides.

Statistics, Metaphysics and Capitalism

Underlying these changes in economy are advances in statistics, algorithmic logic and computation. Society is becoming increasingly algorithmic, and thus governed more and more by statistics, continuing a long trend of the growing use of quantification and probabilistic analysis in the management of populations, economies and life itself.[14] In this new paradigm, scientists and marketers look not to causality to predict how individuals, groups and physical systems will act, but rather to correlations.[15] To paraphrase Karl

---

13 Along similar lines as this book, Scott Timcke has recently argued that algorithms and the inequities they perpetuate must be understood not as a momentary epistemic or democratic crisis, but rather as an intensification and automation of capitalist forces that further forecloses the possibility of politics to address these issues.

14 Louise Amoore, *The Politics of Possibility: Risk and Security beyond Probability* (Durham, NC: Duke University Press, 2013); Dan Bouk, *How Our Days Became Numbered: Risk and the Rise of the Statistical Individual* (Chicago: University of Chicago Press, 2015); Jacqueline Wernimont, *Numbered Lives: Life and Death in Quantum Media* (Cambridge, MA: MIT Press, 2019); Robin James, *The Sonic Episteme: Acoustic Resonance, Neoliberalism, and Biopolitics* (Durham, NC: Duke University Press, 2019).

15 Theodora Dryer, "Algorithms under the Reign of Probability," *IEEE Annals of the History of Computing* 1 (2018): 93–96; Orit Halpern, *Beautiful Data: A History of Vision and Reason since 1945* (Durham, NC: Duke University Press, 2015), 36.

Marx's comments from *The Eighteenth Brumaire*, quoted at the top of this introduction: as far as capital is concerned, it is now through statistics that dead generations, or at least past events, weigh on the brains of the living. To understand contemporary capitalism, one must have some understanding of statistics, and to grasp statistics, one must account for how this particular science is integrated into capitalist production.

There is a constant risk that technology, algorithms and digital systems are given too much agency and seen as the necessary target of political energies—a misapprehension that occludes the ways in which they represent, mediate and accelerate larger social antagonisms. Moreover, we must elucidate the ways in which these forms of social domination give objective force to those systems; that is, we must explain why it is that they are believed in the first place and given the agency they currently enjoy.

This book will argue that it is in statistics and probability that the metaphysical force of these systems is ultimately located. So while this is, in a sense, a text about algorithms, digital culture and technology, its focus is significantly narrower. The following chapters attempt to bring to light the central role that statistics and theories of probability play in shaping the possibilities of these systems and tying them fundamentally to the functioning of capitalism.

In this light, it is important to note that statistics—the field that supplies both the methods and interpretive framework for machine learning and algorithmic society writ large—operates on two distinct levels: metaphysical and mathematical. On the metaphysical level, different schools of statistics make diverse philosophical claims about the relationship between chance, induction and knowledge: some declare probability to be a subjective measure of belief, while others see it as an objective frequency of a collection of outcomes.[16] On the mathematical level, statisticians have built atop

---

16 Lorraine Daston provides an exceptional history of how interpretations of probability split into objective and subjective theories: "How Probabilities Came to Be Objective and Subjective," *Historia Mathematica* 21, No. 3 (1994): 330–344.

these systems a battery of mathematical tests that evaluate how this metaphysical relationship plays out according to an experiment or observed data. Ronald Fisher, one of the founders of modern statistical methods and experimental design, was explicit about the value of this distinction between technical questions and "principles"—or what could be called metaphysical questions:

> The questions involved can be dissociated from all that is strictly technical in the statistician's craft, and, *when so detached*, are questions only of the right use of human reasoning powers, with which all intelligent people, who hope to be intelligible, are equally concerned, and on which the statistician, as such, speaks with no special authority. The statistician cannot excuse himself from the duty of getting his head clear on the principles of scientific inference, but equally no other thinking man can avoid a like obligation.[17]

It is precisely on this metaphysical level that it is possible to understand how revolutions in statistical epistemology, just like other technologies and scientific techniques, have been revolutions in production. These metaphysical questions determine how data can be turned into knowledge and action. Though math often presents itself as a closed system that is relatively stable, to some even a universal language, when it touches on the actually existing world, this relationship becomes open, socially contingent and in flux. This unstable relationship between world and number is especially apparent in the realm of statistics and probability, where these shifts both determine and are determined by social and economic formations.[18]

---

17 Ronald Fisher, *The Design of Experiments* (New York: Hafner Press, 1971), 1–2.

18 Wendy Chun argues that science, and especially models of future events such as global warming, "trouble the separation of science from politics, model from evidence, but also, and more importantly, the normal and normative relationship between understanding and agency": "On Hypo-real Models or Global Climate Change: A Challenge for the Humanities," *Critical Inquiry* 41, No. 3 (2015): 675–703. See also Alain Desrosières, "How Real Are Statistics? Four

Just as Taylorist scientific management revolutionized industrial production at the turn of the twentieth century, the "inference revolution" has revolutionized the production of knowledge and abstraction from data in the late twentieth and early twenty-first.[19] Statistical methods have become central to the functioning of modern capitalism by producing profitable reserves of knowledge via measures of inference: they have turned manufacturing data into a revolution of just-in-time logistics, given rise to high-frequency trading, and financed the modern internet with targeted advertising.

Concurrent with statistics' ability to revolutionize production, they also play a metaphysical role in capitalism, providing a science and method for turning knowledge into action. With this ability, they have been granted a certain force, a power to make things appear as necessary and objective, which they lend daily to the algorithmic systems that control much of contemporary life. The chapters that follow are an attempt to account for this force that appears to be the non-locatable force of reason itself—one that has the ability to compel individuals to labor and live in distinctly different ways but increasingly at the behest of probabilistic systems that seek to maximize profits for capital.

The Ideal Coin

Capitalism and statistics are much more than accidental partners in this digital economy. And nowhere is this metaphysical mutual

---

Possible Attitudes," *Social Research* (2001): 339–355; and, in regard to scientific inference and statistics, Gerd Gigerenzer and Julian N. Marewski, "Surrogate Science: The Idol of a Universal Method for Scientific Inference," *Journal of Management* 41, No. 2 (2015): 421–440. See also Lorraine Daston, "Fitting Numbers to the World: The Case of Probability Theory," in *History and Philosophy of Modern Mathematics*, William Aspray and Philip Kitcher, eds., *Minnesota Studies in the Philosophy of Science*, Vol. 11 (Minneapolis: University of Minnesota Press, 1988), 221–237.

19 Gerd Gigerenzer and David J. Murray, *Cognition as Intuitive Statistics* (London: Psychology Press, 2015).

implication more clear or succinct than in their use of an "ideal coin" to relate individual workers, commodities and data to universal principles. In capitalism, this coin takes form as money, serving as a direct representative of value, always and everywhere exchangeable for its equivalent in terms of a commodity or other coins. If the real, physical coin wears away so much that the value of its metal noticeably decreases, its guarantor (the state) must be willing to replace it with a new one in order to maintain its ideal value.

Likewise, for frequentist statistics—an approach whose specifics we will return to at length—the ideal coin, whose individual flips have no impact on each other, serves as a conceptual support for probability, the measure that lays the foundation for statistical thought and its ability to relate data to universal inference. This timeless and ideal coin allows one to imagine a near-infinite series of flips, whose frequency of heads stays constant and thus becomes the probability of such an outcome. According to frequentist thought, probability merely represents the long-run frequency of such an imaginary instrument.

Neither the universal equivalent of money in capitalism nor the ideal coin of frequentist statistics exists in this world. Rather, like mystical, imagined actors, both types of coins operate in the nether, suturing metaphysical categories of value and probability to the material world. Money allows us to know the world through price. An ideal coin allows us to calculate probability through an ideal of repeatability. Both are what Sohn-Rethel calls "real abstractions" that facilitate abstract and productive ways of understanding the world according to each coin's ideal contours; these are ideal, but also real, being so close at hand, represented by actually existing coins.[20] Frequentist statistics and capitalism successfully generated a regime of knowledge based on this imaginary invention that allows us to think of the world as objective.

While the early and mid twentieth century saw attempts to find an objective ground for statistics, over the last few decades this

---

20 Sohn-Rethel, *Intellectual and Manual Labour*, 20–21.

objective view of statistics has waned as proponents of so-called Bayesian approaches have abandoned the objective terrain established by an ideal coin in favor of a subjective measure—one that starts its calculations with a guess and then persistently updates probabilities as new evidence is gathered. In a Bayesian world, there is no ideal object, but only the experience of a subject who collects data. As we will see, Bayesian methods are ascendant, and the machine learning models built upon them are now used to regulate everything from high-frequency trading to global supply chains.[21]

This movement from objective to subjective foundations for knowledge is revolutionizing the metaphysical production of the world—shifting the very foundations of scientific knowledge—and with it the material production of informational capitalism. It is a transformation of knowledge into a form that is more fluid, local and given over to market dynamics. Yet, this subjectification of knowledge does not overturn the force of objectification, nor does it negate the power of its real abstractions; rather, it provides it a specifically capitalist form that does away with any solid ground or locatable origin, replacing the scientist and her desire for knowledge with the demands of the market.

In this revolution, statistics and capitalism are intimately related in their metaphysical goals and underlying work: both attempt to transform individual data into universal laws. Whether this work

---

21 While machine learning techniques are not exclusively Bayesian in their approach, the Bayesian revolution both opened the door for advances in machine learning and continually informs its developments. Thus, Bayesianism can be seen as a spark that has led to a much-broader revolution in how meaning is extracted from data and updated as new evidence is discovered. See Jon Williamson, "The Philosophy of Science and Its Relation to Machine Learning," in *Scientific Data Mining and Knowledge Discovery: Principles and Foundations*, Mohamed M. Gaber, ed. (New York and Berlin: Springer, 2009), 77–89. There are a multitude of interpretations of Bayesian statistics; our focus in this book is largely on the subjective interpretation developed by de Finetti and Savage. See Bruno de Finetti, "Probabilism: A Critical Essay on the Theory of Probability and on the Value of Science," *Erkenntnis* 31 (1989): 169–223; and Leonard J. Savage, *The Foundations of Statistics*, 2nd ed. (New York: Dover, 1972).

is carried out for the sake of prediction of future stock returns or the creation of a market-based economy, both statistics and capitalism magically transform the local, contingent and individual into the general, universal and global. This magic trick, while clearly material, real and productive, is still metaphysical and can never be fully reduced to mere physicality.[22] Statistics and algorithms, like all knowledge, are founded on an objectified set of beliefs about what is equivalent—or, more accurately, what is made computable through probability, ratio or equivalence.

The heterodox Marxist philosopher Moishe Postone says of capitalism:

> The result is a historically new form of social domination—one that subjects people to impersonal, increasingly rationalised, structural imperatives and constraints that cannot adequately be grasped in terms of class domination, or, more generally, in terms of the concrete domination of social groupings or of institutional agencies of the state and/or the economy. It has no determinate locus and, although constituted by determinate forms of social practice, appears not to be social at all.[23]

These real abstractions, as measures both of the economic value that labor produces and the scientific value of a hypothesis, act as a marker of this form of rationalization and its lack of material locus. One can believe or hope however they like, but at the end of the day, it is by the dictates of this objectified force that both capitalism and statistics measure our success, and hence our ability to survive. This non-locatable impulse, both for the production of knowledge and value, is what constitutes what Postone calls "abstract domination." It is through these non-locatable vectors of

---

22 Alberto Toscano, "Materialism without Matter: Abstraction, Absence, and Social Form," *Textual Practice* 28, No. 7 (2014): 1221–1240.

23 Moishe Postone, "Critique and Historical Transformation," *Historical Materialism* 12, No. 3 (2004): 59.

abstract domination and control that algorithmic systems and statistics shape the world today.

This does not mean that capitalism or statistics do away with prior concrete domination based on race or gender; rather they use it to justify such domination in terms of abstract vectors and maintain them in concrete form. For instance, an individual's inability to secure a mortgage or insurance can be used to create segregation in the form of redlining, concrete domination that is further enforced through targeted police violence. Scholar and professor of African and African diaspora studies, Simone Browne, in detailing the origins of modern surveillance in the history of the enslavement of Black people, demonstrates how capitalism and technologies of control do not escape the violent and racist histories that produce them, and instead tend to reproduce and reconfigure the regimes of legibility from which they are born. Browne argues that when "technologies, in their development and design, leave out some subjects and communities for optimum usage, this leaves open the possibility of reproducing existing inequalities."[24]

Thus, while algorithms and machine learning may change the speed and nature of computation, they are ultimately bound to reproduce extant societal systems of valuation and violence. So, if we are to resist the inequities of algorithmic logic, it cannot be only at the level of technology, nor exclusively at the level of metaphysics. Abstract and concrete forms of domination interact and reinforce each other, just as classed, racialized, gendered and imperialist oppression and violence interact.[25] However, while efforts to resist and rearrange the forces of capitalist and statistical metaphysics can open new possibilities, such a shift alone will solve nothing—and is likely impossible.

---

24 Simone Browne, *Dark Matters: On the Surveillance of Blackness* (Durham, NC: Duke University Press, 2015).

25 Kimberle Crenshaw, "Mapping the Margins: Intersectionality, Identity Politics, and Violence against Women of Color," *Stanford Law Review* 43 (1990): 1241; Combahee River Collective, *A Black Feminist Statement* (1977).

Toward a New Objectification

More than ever, it is the very concept of the political subject that is defeated by digital technologies: hope for a "Twitter revolution" has given way to accusations of political meddling by foreign powers, and the promise of free and accessible information has been replaced with government surveillance and censorship.[26] Since Marx, if not earlier, radical political theory has tended to rely upon the existence (or creation) of a revolutionary subject that would use some combination of will, position and knowledge alongside a certain force of history to overthrow extant forms of power and oppression. But today, access to transformative politics founded on a voluntaristic subject seems increasingly unfathomable. Industry lobbyists with carefully crafted narratives, a twenty-four-hour news cycle that passes too quickly to allow organized political response, targeted advertisements, and the ease with which corporations move production elsewhere at the hint of new labor demands are all tactical threats to the prospect of any contemporary radical political project.

In traditional Marxist theory, the proletariat as a class constitutes this revolutionary subject. To paraphrase the last lines of *The Communist Manifesto*, workers have only to unite to become a collective revolutionary subject. While there have been endless debates around the specific nature of the revolutionary subject, the central role of this subject to revolutionary thinking is beyond doubt.[27]

But despite the centrality of this dream of an efficacious political subject, the subject of radical political theory, especially in Western Enlightenment-inspired varieties of communism, has

---

26 Gilles Châtelet provides an excellent and damning synopsis of the naturalization of markets as a means to manage society in *To Live and Think Like Pigs: The Incitement of Envy and Boredom in Market Democracies*, trans. Robin Mackay (New York: Urbanomic, 2014).

27 Consider, for example, the question brought to the fore by Vladimir Lenin and the Russian Revolution on whether this subject must begin its work in the industrialized capitalist countries or an agricultural country like Russia; or the debates about whether the revolutionary subject chooses revolution as an act of will or is, rather, swept along by the forces of history and economics.

been in crisis since at least the 1960s, if not from its very origins, which were themselves often an attempt to universalize a very specific white male British subject in the form of the proletariat.[28] The struggles of student radicalism in the late '60s; the emergence of non-class-based liberal movements (e.g., the environmental movement, anti-war movement and human rights organizations); the recognition that this ideal of a universal subject was likely just a Western white male fantasy; the realization of the violent excesses of Stalinism and Maoism; the collapse of the Soviet Union; and the "return" of ethno-national conflict and religious fundamentalism have all served to call into question the viability of a unified political subject who could foment global revolutionary change.

The rise of algorithmic systems that isolate individuals within their own unique "filter bubble" threatens to further mitigate the possibility of an effective political subject.[29] Today, the revolutionary subject is beset simultaneously by an algorithmically fragmented reality and an intensely managed digital control. While the former modulates individuals' reality by data-driven algorithms that personalize the news they read, prices for products, and what healthcare they have access to, the latter appears everywhere that digital systems are used for control: it regulates workers in the factory, students in the university, prisoners (even after their release), and immigrants both between and within countries.

Many scholars and activists have detailed how these technologies control individuals and reinforce existing inequalities, yet it seems increasingly unlikely that we will overcome the larger political and economic forces that operationalize these technologies.[30]

---

28 Cedric Robinson, *Black Marxism: The Making of the Black Radical Tradition* (Chapel Hill: University of North Carolina Press, 2000).

29 Eli Pariser, *The Filter Bubble: What the Internet Is Hiding from You* (New York: Penguin, 2011).

30 See Seb Franklin, *Control: Digitality as Cultural Logic* (Cambridge, MA: MIT Press, 2015); Alexander Galloway, *Protocol: How Control Exists after Decentralization* (Cambridge, MA: MIT Press, 2004); Virginia Eubanks, *Automating Inequality: How High-Tech Tools Profile, Police, and Punish the Poor* (New York: St. Martin's Press, 2017); Safiya Umoja Noble, *Algorithms of Oppression: How Search Engines Reinforce*

The Western Marxist and revolutionary dream of dereification—a formula of Georg Lukács wherein the task of individuals was to understand themselves and history—has given way to deep and prevailing cynicism.[31] No solidarity seems to hold together, and the only possible progressive politics appear as mere hospice for our chronic ecological and social ills. While the causes of the fracturing of a universal political subject are many—and its exclusions and shortsightedness likely mean such a project was never feasible—the ability for technology to modulate, divide and distract offers both a metonymy and a ground to theorize this much-larger problem.

There are many who argue that creative power, or force of will, can produce an inviolate subject who stands outside the cold, calculating power of computation. From Tiqqun's resistance against the "cybernetic hypothesis," to Franco "Bifo" Berardi's celebration of the "insolvency of language," to Jodi Dean's return of the party, numerous theoretical and philosophical projects have attempted to rescue some sort of subject able to resist the ravages of modern capitalism.[32] It is not my intention to disparage these

---

Racism (New York: New York University Press, 2018); and Siva Vaidhyanthan, *The Googlization of Everything (And Why We Should Worry)* (Berkeley: University of California Press, 2011).

31 Georg Lukács, *History and Class Consciousness: Studies in Marxist Dialectics*, trans. Rodney Livingstone (Cambridge, MA: MIT Press, 1967); Axel Honneth, *Reification: A New Look at an Old Idea* (Oxford: Oxford University Press, 2008); Peter Sloterdijk, *Critique of Cynical Reason* (Minneapolis: University of Minnesota Press, 1988).

32 Tiqqun, "L'Hypothèse cybernétique," *Tiqqun* 2 (2001): 40–83; Franco "Bifo" Berardi, *The Uprising: On Poetry and Finance* (Los Angeles: Semiotext(e), 2012); Jodi Dean, *Crowds and Party* (London and New York: Verso, 2016). Likewise, Brian Massumi argues that "the first task of the revaluation of value is to uncouple value from quantification. Value must be recognized for what it is: irreducibly qualitative" (Thesis 5). While Massumi admirably turns against the fetishization of the capitalist subject and toward the reconceptualization of value, his disavowal of quantification (and with it computation) risks resulting, in the end, in an inability to overcome the distributed computation that is capitalism. *99 Theses on the Revaluation of Value: A Postcapitalist Manifesto* (Minneapolis: University of Minnesota Press, 2018). See also Seb Franklin's calls in *Control* for resistance founded on "states of undecidability or unmeasurability."

endeavors; on the contrary, I hope they succeed. Yet there is a serious threat that the forces of oppression confronted by humanity are all too capable of absorbing and managing the disturbances such creative, uncountable subjects portend.

In this vein, the communist collective Endnotes summarizes the situation well, suggesting that no unifying force currently exists to hold together a revolutionary political coalition: "At present there seems to be no class fraction—whether 'the most strategically placed' or 'the most oppressed'—whose struggles express a general interest. At the same time, attempts to conjure up a new unity from this diversity by simply renaming it as 'multitude' or 'precariat,' for example, merely gloss over this fundamental problem of internal division."[33] In sum, the contemporary landscape of radical politics is so fragmented—and its potential subjects so divested from the "general interest" of collective action—that no group could operably achieve what was expected of the proletariat in classic Marxist theory. Moreover, as Postone has argued, "attempts to rescue human agency that posit historical contingency abstractly and transhistorically, bracket and veil the existence of historically specific structures of domination. They are thereby, ironically, profoundly disempowering."[34] The very structure of this subject appears to guarantee its undoing. Today, the revolutionary subject is under siege and in doubt. Perhaps the psychoanalyst Jacques Lacan put this doubt most cruelly and succinctly when he said to the French students of May '68: "What you aspire to as revolutionaries is a master. You will get one."[35]

In the face of the subject's foreclosure, others have abandoned this subjective pursuit in favor of an objective ground for politics.[36]

---

33 Endnotes Collective, *Endnotes 4* (London: Endnotes, October 2015).

34 Postone, "Critique and Historical Transformation," 56.

35 Jacques Lacan, *The Seminar of Jacques Lacan: The Other Side of Psychoanalysis*, Book XVII, trans. Russell Grigg (New York: W.W. Norton & Company, 2007), 207.

36 See, for example, Bruno Latour, "Why Has Critique Run Out of Steam? From Matters of Fact to Matters of Concern," *Critical Inquiry* 30, No. 2 (2004):

Yet while many well-meaning individuals hope to reaffirm the solidity of the world and facts atop the bedrock of scientific belief, this objective position tends, in its desire for facts, to present the conditions of the capitalist world as humanity's only option.[37] In turn, its objective ground ends up depoliticized or, worse yet, used to reinforce current white-supremacist and capitalist understandings of the world. All too often, individuals seeking technical solutions to social problems reproduce the very systems they attempt to work against.[38]

Stuck between these two poles of what we might—at the risk of misusing the vocabulary of dialectic materialism—call "objective" and "subjective" leftism, the most politically efficacious path forward may be neither a directly hybrid route nor a privileging of one specific pole. Instead, we must work directly on the torsion between the objective and the subjective. Marx tells us, especially in *Capital*, that capitalism, and with it politics under capitalism, is founded upon a metaphysical process wherein humanity's social and economic world—the world of value and wage labor—is twisted and presented as the objective, unchangeable state of things. While the entire production process is social, and hence an agreement between subjects—in essence subjective—it does not matter what we think or desire; if we do not own capital, we must sell our labor on the market at the going rate. Philosophy be damned, labor is worth what the market is willing to pay.

Marx gives the name "objectification" to this metaphysical process, by which socially negotiated relationships between people are made to appear as relations between objects, and hence

---

225–248; Graham Harman, *The Quadruple Object* (London: Zero Books, 2011); Ian Bogost, *Alien Phenomenology; or What It's Like to Be a Thing* (Minneapolis: University of Minnesota Press, 2012); Levi Bryant, *The Democracy of Objects* (Ann Arbor, MI: Open Humanities Press, 2011).

37 Mark Fisher, *Capitalist Realism: Is There No Alternative?* (London: Zero Books, 2009).

38 Evgeny Morozov, *To Save Everything, Click Here: The Folly of Technological Solutionism* (New York: Public Affairs, 2013).

as objective. Instead of focusing on what is *objectively true*—the tactic of the above objective leftism—Marx's theory of objectification allows us to interrogate the process by which certain things appear objective: the moment when decisions of social actors cease to appear as decisions and begin to appear as natural and unimpeachable.

Let us be clear: while the two are intimately related, the history and future of objectification is not the history of objectivity. Those histories have been well written elsewhere.[39] Our primary concern is with the way that objects come to account for human affairs, managing, recording and presenting our social and intersubjective interactions as part of a larger framework of rationality and objectivity, and with the force this management is given. Objectification is, in short, the way in which social production is inserted into the larger history of scientific (or, prior to that, theological) objectivity.[40]

To understand the nature of contemporary capitalism—specifically its ability to present a complex system of social exploitation as objective and natural—it is imperative to analyze how statistics and the algorithmics of machine learning take up and continue the commodity's legacy of mediating society and economics "behind our backs." It is no longer enough to simply comment on the importance of data and computing to capitalism. Nor will it suffice to merely explain the mathematics behind relevant algorithms and excavate the sets of data used to train them. Instead, these algorithms must be understood and addressed as systems that take up the social work of objectification, thinking for us and managing our affairs *objectively*. Just like the commodity and its value,

---

39 See Peter Galison and Lorraine Daston, *Objectivity* (New York: Zone Books, 2007); Theodore M. Porter, *Trust in Numbers: The Pursuit of Objectivity in Science and Public Life* (Princeton, NJ: Princeton University Press, 1995); and Michel Foucault, *The Order of Things* (New York: Routledge, 2005, repr.).

40 Antonio Negri analyzes the relationship between the subjective and objective sides of capitalism at length, especially in relation to Marx's writings in the *Grundrisse*, tracing the ways in which they interact and mutually constitute each other. Antonio Negri and Jim Fleming, *Marx beyond Marx: Lessons on the Grundrisse* (Brooklyn: Autonomedia, 1991).

disbelief in them does not make them any less powerful; and, this fact is what must be accounted for.

We sit at a moment where the monstrous objects of these algorithms matter a great deal. Today, it is precisely computation—from climate change models to the use of machine learning to provide personalized shopping, news and dating, as well as distribute state and corporate violence—that determines our current objective social reality under digital capitalism. As algorithms and data increasingly produce both scientific knowledge and commercial value, taking on tasks ranging from the distribution and extraction of value in high-frequency trading to the management of just-in-time supply chains to the "disruption" of traditional ossified industries, old forms of material and metaphysical production are rent asunder.

Now algorithms are built on statistics that leverage and produce this force of objectification, tying data to decisions and suturing the particular to the universal, the subject to the object, and the material to the metaphysical. Again, the work that these algorithms do—work that is fundamentally built on advances in statistics and probability—is simultaneously material and deeply metaphysical. While they use magnetic bits for practical purposes, calculating how to move capital and physical goods around the world, they also redefine the world in uniquely abstract ways, turning data about that world into concepts, predictions and inferences. These freshly minted, algorithmically produced abstractions then become the new indices of scientific and commercial knowledge.

In this light, a revolutionary theory that works on objectification, rather than on subjects or objects, must focus directly on the relationship between computation and the production of knowledge and value. In order to account for these objectified forces that determine political possibilities, we must confront not merely the economic processes governing these relationships but the structures of objectified belief that allow computational systems to work behind our backs. We seek, then, a theory of the revolutionary object—or, more precisely, a theory of a revolutionary objectification, of how it may be possible, in the absence of a single, unified subject, to revolutionize social relations. In short, we are looking for a way to revolutionize "what counts" without making recourse to either an imagined totality or a transhistorical subject.

To be clear, the adoption of such a position does not entail a wholesale opposition to math, statistics or science. Quite the opposite: these are necessary components of radical politics and thought in the twenty-first century. "A post-statistical society," writes political economist William Davies, "is a potentially frightening proposition, not because it would lack any forms of truth or expertise altogether, but because it would drastically privatise them."[41] To simply refuse statistics as a capitalist tool would leave humanity subjected, all the more so, to these mathematical decisions.

While decades of Marxist thought have been dedicated to the concept of "dereification," or revealing what is "truly" happening in these objectified relations, another path is possible. We must start from the fundamental impossibility of some final revelation about the ultimate nature of complex algorithmic models, or about the opaque nature of proprietary systems like capitalism. We should refuse the simple assumption that objectification—and thus abstraction itself—is necessarily bad or unacceptably reductive. As a process, objectification lies at the heart of the very nature of all technology: from the moment a stream of water touched the buckets of the first waterwheel, the forces of nature came to be objectified and to objectify human thinking about the world. Humanity, now blessed with waterpower, had moved the baseline of possible thought from a river and its currents to the mill and its axis. In the centuries since, such abstractions about the world and its potential have run from the undomesticated kinetic energy of the river to the rivers of plasma that flow through experimental nuclear fusion reactors.

---

41 William Davies, "The Long Read: How Statistics Lost Their Power—And Why We Should Fear What Comes Next," *Guardian*, January 19, 2017.

After centuries of advances in thermodynamic abstraction, statistically based technologies like machine learning now form the material base of objectification in our current knowledge-driven economy, where data of our economic, political and social worlds are reflected back to us as objective measures: the stock market is up; here is the coat you are likely to buy; and so on. The knowledge and abstractions these technologies produce are not wrong or incorrect, any more than the price of corn at market is wrong. Rather, we must understand how algorithmic "objects" function, what social relations they make objective, what form this objectivity takes, and the possibilities of new objectifications that lie in their contradictions. We do not yet know what form these new objectifications could take, but—much like the commodity form reshaped the world—there is revolutionary value in seeking out these new objectifications; current debates, contradictions and advances in statistics and machine learning are likely to play a central role in shaping these new forms. Thus, the point is not one of dereification—to show how things really and truly are, or to get at the secret of a metaphor or an analogy—but rather to show how algorithmic objects work and the displacements they effect.[42]

Political Economy

Here, there emerges an immensely important set of questions that concern the means through which the global economy functions, as well as the extent to which digital systems have fundamentally altered the function of that economy. While these questions take

---

42 Slavoj Žižek argues that this is how the symptom functions in both Marx and Freud. The aim is not to get behind the mask, because one only discovers that there is nothing there; rather one must understand how the displacement translates material into meaning—labor into value in the case of Marx and the unconscious thought into the dream symbol for Freud. What matters is the form of translation, not the meaning of the symbol. *The Sublime Object of Ideology* (London: Verso, 1989).

on a multiplicity of valences and have been addressed in a variety of fields, including in literature on digital culture and political economy, three elements stand out that bear directly on this work. The first is an ongoing debate about the existence and nature of "free labor" on digital platforms, and more generally about various forms of immaterial labor that work on data, information and knowledge rather than on the production of commodities.[43] In this regard, it is especially important to remember that these forms of labor, often thought of as digital, are part of a much-larger,

---

43 Christian Fuchs and Sebastian Sevignani have updated Dallas Smythe's concept of the audience commodity to clarify the process of productive consumption: "What Is Digital Labour? What Is Digital Work? What's Their Difference? And Why Do These Questions Matter for Understanding Social Media?," *tripleC: Communication, Capitalism and Critique* 11, No. 2 (2013): 237–293; "Digital Prosumption Labour on Social Media in the Context of the Capitalist Regime of Time," *Time and Society* 23, No. 1 (2014): 97–123. Lisa Nakamura has used online video games to show how even platforms built around social play can create the conditions for the production of digital commodities that can be sold to others, exploiting labor along racialized and imperial demarcations: "Don't Hate the Player, Hate the Game: The Racialization of Labor in World of Warcraft," *Critical Studies in Media Communication* 26, No. 2 (2009): 128–144. Jonathan Beller has argued that the cinematic image and its legatees have shaped contemporary production with spectators producing value: *The Cinematic Mode of Production: Attention Economy and the Society of the Spectacle* (Lebanon: Dartmouth College Press, 2006). Tiziana Terranova, drawing on the insights of Italian autonomism, has proposed that most internet users are everywhere and always producing without necessarily receiving compensation, further arguing that to call unpaid digital work "labor" has a political value that exists even beyond its economic accuracy: "Free Labor: Producing Culture for the Digital Economy," *Social Text* 63, No. 18 (2000): 33–58; "Free Labor," in *Digital Labor: The Internet as Play-Ground and Factory*," in Trebor Scholz, ed. (Abingdon: Routledge, 2013). See also Lisa Nakamura, *Digitizing Race: Visual Cultures of the Internet* (Minneapolis: University of Minnesota Press, 2008); Dallas Smythe, *Dependency Road: Communications, Capitalism, Consciousness, and Canada* (Norwood, NJ: Ablex Publishing Corporation, 1981); Christian Fuchs, *Digital Labour and Karl Marx* (London: Routledge, 2014); Nick Dyer-Witheford, *Cyber Marx: Cycles and Circuits of Struggle in High-Technology Capitalism* (Urbana, IL: University of Illinois Press, 1999). Edward Conor, "Revisiting Marx's Value Theory: A Critical Response to Analyses of Digital Prosumption," *The Information Society* 31, No. 1 (2015): 13–19.

evolving capitalist system; for example, understanding physical hardware must include the energy it requires, the extraction of raw materials and the continued reliance on dangerous and exploitative industrial production, especially in the global South.[44]

A second, related question concerns the nature of value extraction: that is, precisely how it is that digital systems produce, extract, valorize or aid in the production of value. Building upon the theoretical legacy of Italian autonomism and *operaismo* (workerism), a succession of scholars have greatly expanded the definition of value production to include "immaterial labor" by utilizing the concept of the social factory, wherein all sorts of non-waged labor still produce value for society.[45] Much of this work has taken inspiration from Marx's claim, in his famous "Fragment on the Machine," that as the "general intellect" develops, it allows increasingly self-productive machines to produce value directly.[46]

Third, a substantial amount of work on the rise of data and information technology has attempted to periodize the history of

---

44 Nick Dyer-Witheford, *Cyber-Proletariat: Global Labour in the Digital Vortex* (London: Pluto Press, 2015). Jussi Parika, *A Geology of Media* (Minneapolis: University of Minnesota Press, 2015). Christian Fuchs, *Digital Labour and Karl Marx* (London: Routledge, 2014); Nikhil Pal Singh, "On Race, Violence, and So-Called Primitive Accumulation," *Social Text* 34, No. 3 (128) (2016): 27–50; and Intan Suwandi, *Value Chains: The New Economic Imperialism* (New York: Monthly Review Press, 2019). Saskia Sassen has also traced the myriad ways in which contemporary capitalism seeks to expel individuals from the system of capitalist production: *Expulsions: Brutality and Complexity in the Global Economy* (Cambridge, MA: Harvard University Press, 2014).

45 Michael Hardt and Antonio Negri, *Empire* (Cambridge, MA: Harvard University Press, 2000).

46 Others, such as Nick Srnicek and Jathan Sadowski, have argued that data should be considered closer to a raw material from which data can be extracted or, in a related vein, that digital platforms can be considered property from which rent can be extracted. Nick Srnicek, *Platform Capitalism* (New York: Polity, 2017); Jathan Sadowski, "When Data Is Capital: Datafication, Accumulation, and Extraction," *Big Data and Society* 6, No. 1 (2019). David Harvey argues that for Marx, the value theory of labor is not static but is instead subject to change as capitalism transforms itself : "Marx's Refusal of the Labour Theory of Value," *Reading Marx's Capital with David Harvey*, March 1, 2018.

capitalism and global society in general, and to explain the ways in which digital capitalism differs from industrial capitalism. A range of theorists have argued for varying degrees of a "break" or transition from earlier forms of capitalism. They have variously referred to this new form as surveillance capitalism, cognitive capitalism, network society or platform capitalism. Some identify the shift as a control revolution, while others frame it as the next industrial revolution or a data revolution; still others contend, on the contrary, that there has been no major break in the history of production at all.[47]

While all three of these debates around the nature of twentieth- and twenty-first-century political economy are of immense importance, to address them directly and systematically would require a fundamental rethinking of all three volumes of Marx's *Capital* and a synthesis of a vast amount of work that has followed since their original writing. Such a project would likely require at least as many volumes, if not more. Thus, this book is not so much a description of the exact nature of contemporary political economy at this precise moment as it is an argument for the importance of political economy; it is an attempt to show the depths to which these questions of political economy shape the very production of knowledge and the ways in which knowledge production shapes political economy. While certain commitments and understandings will likely be noticeable throughout, it

---

47 See Shoshana Zuboff, *The Age of Surveillance Capitalism: The Fight for a Human Future at the New Frontier of Power* (London: Profile Books, 2019); Yann Moulier-Boutang, *Cognitive Capitalism* (Cambridge, UK: Polity, 2011); Manuel Castells, *The Rise of the Network Society* (Hoboken, NJ: John Wiley & Sons, 2011); Srnicek, *Platform Capitalism*; James Beniger, *The Control Revolution: Technological and Economic Origins of the Information Society* (Cambridge, MA: Harvard University Press, 2009); Klaus Schwab, *The Fourth Industrial Revolution* (New York: Crown Business, 2017); Viktor Mayer-Schönberger and Kenneth Cukier, *Big Data: A Revolution That Will Transform How We Live, Work and Think* (New York: Houghton Mifflin Harcourt, 2013); and Sam Popowich, "Mechanical Animals: Big Data, Class Composition, and the Multitude," University of Alberta Libraries, 2019.

is hoped that regardless of how one understands modern political economy to function, what follows will make apparent the importance of these larger questions to the very ground of what constitutes knowledge today.

It is clear, especially given that machines now produce knowledge directly, that political economy does not simply take up already-existent technologies and put them to use, or even develop them; rather, these technologies are themselves techniques for shaping and changing political economy. Thus, as Postone argues, the very shape of political economy is itself plastic and changed by historical circumstances: "Inasmuch as Marx analyses social objectivity and subjectivity as related intrinsically, this focus on the historical specificity of his categories reflexively implies the historical specificity of his theory. No theory, within this conceptual framework, has transhistorical validity. Rather, the standpoint of critical theory must be intrinsic to its object."[48] In sum, even if it changes rapidly or over centuries, it is the very nature of political economy that is at stake in the process of objectification. This book aims to trace this relationship in terms of the ways it is changed by statistics, opening questions of political economy rather than attempting to answer them definitively.

That said, the final chapter of this book will make some arguments that address these questions, with a focus on the development and enclosure of the general intellect. While these arguments are built on implicit understandings of the function of machinery and social labor that align with some of the autonomist readings of digitally mediated production, the larger threat of this enclosure to the production of shared social knowledge remains, whether one accepts this or another account of political economy.

---

48 Postone, "Critique and Historical Transformation," 57. Or, as David Harvey similarly argues: "The formulation of value in the first chapter of *Capital* is revolutionized by what comes later. Value becomes an unstable and perpetually evolving inner connectivity (an internal or dialectical relation) between value as defined in the realm of circulation in the market and value as constantly being redefined through revolutions in the realm of production" ("Marx's Refusal").

Thus, regardless of whether automation produces value directly or merely drives this "treadmill dynamic," as Postone calls it, statistics, probability and automation now play an undeniably important role in contemporary capitalism, one that as we shall see ties the production of knowledge to the exchange of value. Information and media scholar Nick Dyer-Witheford lays out the stakes well:

> The conjunction of automation and globalization enabled by information technology raises to a new intensity a fundamental dynamic of capitalism—its drive to simultaneously draw people into waged labour and expel them as superfluous un- or underemployed. This "moving contradiction" now manifests as, on the one hand, the encompassing of the global population by networked supply chains and agile production systems, making labour available to capital on a planetary scale, and, on the other, as a drive towards the development of adept automata and algorithmic software that render such labour redundant.[49]

Statistics and probability provide the fundamental mathematical logic of many of these aspects of contemporary capitalism, from targeted advertisements to the management of production, distribution and consumption, as well as the expulsion of labor from these activities.[50]

Moreover, beyond these questions of political economy, there is an extensive and growing critical literature on algorithms and quantification in general, much of which engages in what we could call a biopolitical critique of these technologies, drawing on the work of French philosopher Michel Foucault.[51] While there is

---

49 Dyer-Witheford, *Cyber-Proletariat*, 15.

50 Nick Dyer-Witheford, Atle Mikkola Kjøsen and James Steinhoff have recently written of AI: "Capitalism is the fusion of these technological and social logics and AI is the most recent manifestation of its chimerical merging of computation with commodification." *Inhuman Power: Artificial Intelligence and the Future of Capitalism* (London: Pluto Press, 2019).

51 John Cheney-Lippold, *We Are Data: Algorithms and the Making of Our Digital Selves* (New York: New York University Press, 2018); Galloway, *Protocol*.

significant nuance to the argument of and within this critique, it could be summarized thus: starting with early European interest in demography, public health, police, schooling and prisons, modern states have attempted to manage populations by quantifying them and measuring them against some idea of what is normal. So, deviancy in the factory, at school, at home, at the border and beyond are measured and thus managed.[52] A number of scholars, such as Simone Browne and Alexander Weheliye, have criticized and built upon these arguments to demonstrate the centrality of racism, and especially the Atlantic slave trade, to the development and deployment of biopolitical control.[53]

Many explorations of algorithms and big data have followed along similar lines, critiquing the ideologies, power dynamics and decision-making processes behind these algorithms. These arguments, ranging from radical criticisms to more liberal calls for regulation, demonstrate the ways in which these systems build upon the longer history of the statistical management of populations.[54] Some of the exemplary work tracing the injustices of algorithmic society, such as that of Safiya Noble, Virginia Eubanks and Cathy O'Neil, tends both to demonstrate the ways in which

---

[52] Specifically in regard to the history of statistics, see Ian Hacking, *The Taming of Chance* (Cambridge, UK: Cambridge University Press, 1990); and Theodore M. Porter, *The Rise of Statistical Thinking, 1820–1900* (Princeton, NJ: Princeton University Press, 1986).

[53] Browne, *Dark Matters*; Alexander Weheliye, *Habeas Viscus: Racializing Assemblages, Biopolitics, and Black Feminist Theories of the Human* (Durham, NC: Duke University Press, 2014). Likewise, authors such as Didier Bigo have argued that in order to understand modern biopolitics, we must account for how they operate at the peripheries of the state, where sovereign force is often much more violent and apparent in attempts to secure its borders than in the metropole. "Globalized (In)security: The Field and the Ban-opticon," in *Terror, Insecurity and Liberty: Illiberal Practices of Liberal Regimes after 9/11*, Didier Bigo and Anastassia Tsoukala, eds. (London: Routledge, 2008), 20–58.

[54] Wendy Hui Kyong Chun, "Queerying Homophily," in Clemens Apprich, Wendy Hui Kyong Chun, Florian Cramer and Hito Steyerl, *Pattern Discrimination* (Lüneburg: Meson Press, 2018), 59–97; Jacqueline Wernimont, *Numbered Lives: Life and Death in Quantum Media* (Cambridge, MA: MIT Press, 2019).

algorithmic society builds upon earlier modes of statistically informed domination and to argue that massive stores of data and computing power have created a new era of individualized manipulation.[55]

These works are powerful and insightful, but this present text aims to take a slightly different course. Rather than exclusively explicate how algorithms work or provide extensive examples of what they have wrought, these chapters explore the metaphysical and economic force through which they claim to turn data from the world into knowledge *about* that world, and the particular sets of actions such processes require. Thus, the book's central questions are less about how statistics function as an ideology, in the classic sense, or how their methods serve as an analogy or organizing principle for neoliberalism—though certainly those are important and relevant questions. Instead, it aims to address how statistics and theories of probability are directly productive, that is, how they tie together the production of knowledge in contemporary capitalism at a metaphysical level. In this sense, its approach is markedly different from the aforementioned texts documenting the twentieth- and twenty-first-century turn toward quantification, which have rarely focused, or even commented, on the shift from frequentist to Bayesian approaches, a shift that has been

---

55 Noble, *Algorithms of Oppression*; Eubanks, *Automating Inequality*; Cathy O'Neil, *Weapons of Math Destruction: How Big Data Increases Inequality and Threatens Democracy* (New York: Broadway Books, 2016). Likewise, a number of authors have shown the ways in which these digital systems and their antecedents have provided an ideological form for contemporary neoliberalism. Wendy Hui Kyong Chun has shown how software provides a framework for neoliberal governance to understand itself; Jean Pierre-Dupuy has argued that cybernetics and the early days of cognitive science have greatly shaped the contemporary understanding of thought and humanity; and Seb Franklin has suggested the ways in which digitality has provided a set of metaphors through which capitalism thinks and shapes the contemporary subject. Wendy Hui Kyong Chun, *Programmed Visions: Software and Memory* (Cambridge, MA: MIT Press, 2011); Jean-Pierre Dupuy, *The Mechanization of the Mind: On the Origins of Cognitive Science* (Princeton, NJ: Princeton University Press, 2000); Franklin, *Control*.

central to the growing use of probability to manage governments and economies.[56]

While there is much to commend in a critique that centers on statistics, probability and their implementation in the form of algorithmic software as an ideology, a cultural or superstructural form, the present text is concerned with the way these technologies function as a "real abstraction," that is, an abstract form that functions on the level of economy and value. The autonomist Marxist Franco Berardi explains money and language in such terms: "Language and money are not at all metaphors, and yet they are immaterial. They are nothing, and yet can do everything: they move, displace, multiply, destroy. They are the soul of Semiocapital."[57] Whether one believes we are in the age of semiocapital, or just capital, the point remains that statistics functions primarily in the same way: not as a metaphor for anything, but rather as an immaterial thing that can do everything. This is not because statistics is fundamentally ideological in some classic sense, nor because it provides a metaphor for thinking neoliberalism (although it does both of these things), but rather because it functions as a machine that turns data into both value and scientific law.[58]

In the shift from frequentist to Bayesian statistics, where statistics move from the discernment of knowledge to the production of action, these mathematical methods function directly on the level of value. They cease to aid in the selection of what is true, pointing instead to which course of action will produce the most profit. In this way, statistics come to be objectifying, in the sense that they

---

56 For a rare exception in this regard, see Gerd Gigerenzer, Zeno Swijtink and Lorraine Daston, *The Empire of Chance: How Probability Changed Science and Everyday Life* (Cambridge, UK: Cambridge University Press, 1990).

57 Franco "Bifo" Berardi, *The Soul at Work: From Alienation to Autonomy* (Los Angeles: Semiotext(e), 2009), 22.

58 To draw on the language of Donald Mackenzie, statistics and probability are certainly not a camera, and to think of them solely as making propaganda movies is not only wrong, but misses the ways in which they are fundamentally an engine that produces value and capitalism itself. *An Engine Not a Camera: How Financial Models Shape Markets* (Cambridge, MA: MIT Press, 2008).

allow one to understand what is "objectively" true: which decisions the market will most likely reward and which it will punish. They come to think for us, revealing our social world—not in so much as it is social, but rather in so much as it is presented back to us as the will of objects. This is not to say that it is nonideological, but rather that its force derives, at least in part, from its ability to function on the level of value directly, on that world of objectified exchange.

Revolutionary Mathematics

This book draws a certain inspiration from George Berkeley's 1734 text *The Analyst; or, A Discourse Addressed to an Infidel Mathematician, wherein It Is Examined whether "The Object," Principles," and "Inferences" of Modern Analysis Are more Distinctly Conceived, or more Evidently Deduced, than "Religious Mysteries" and Points of Faith."* In a tirade against calculus—a story more fully told in Chapter 2—Berkeley decried those who dared to make use of the infinitesimal that was both there and, in its vanishing, not there. For Berkeley, a man of God, the only knowledge was that of God, and was thus grounded by God. For the infidel, knowledge of calculus was groundless, available to be made and remade, entailing concepts built on concepts, and abstractions built on abstractions. While the mathematics underneath the calculation of these infinitesimals in calculus "worked," they were not considered reasonable. According to his Christian-based rationality, they were to be distrusted as ungodly, and those articulating them as infidel.

Commodity exchange and statistical knowledge production share this ability to refashion the world in new, metaphysical forms, finding knowledge and value that exists neither in the materiality of the world, nor in the realm of some Christian heaven. Capitalism "discovers" value in commodities that allow for exchange, refashioning that value as a quantitative rather than qualitative difference. Statistical processing and machine learning calculate correlations, probabilities and ratios, refashioning the world into insights derived from data. It is through a fundamentally metaphysical act of "making computable" that statistics and capitalism objectify the world; it is through making labor time equivalent with itself and economic value, or making data equivalent with other data and ultimately with inferences, that allows these processes to work.

But today, both statistics and capitalism are in crisis—a crisis that derives fundamentally from political economy and its metaphysical support. This crisis therefore demands a revolutionary approach to algorithmic knowledge. To effect a change in the production of our current world will require us to work on both the political and the metaphysical levels of knowledge production. It will require an intervention in the objective production of new forms of equality and hence in computation, and a reconfiguration of the means by which we make the incommensurate commensurate. In short, it will require the work of an infidel mathematics, a revolutionary mathematics, to create new forms of exchangeability that allow for the development of knowledge beyond and after capitalism. Currently, both commodity exchange and statistics equate things, but they do so only in the service of capital accumulation. It is already too late to simply point out the inequity of these calculations; instead, we must rework the metaphysical equatability that undergirds exchange itself—the unreasonable ground of our very reason.

Thus, the work of a revolutionary mathematics is not exclusively or even primarily that of the mathematician or the specialist. Rather than directly technical, the work of this mathematician is metaphysical: their task is to rework the meaning of what is calculated and to figure out how to live and reason under and toward a new computability and different forms of exchange. While such a mathematics must not shy away from technical questions, it must be concerned with the process of reason itself. In this regard, it is nearly impossible to say this or that form of thinking or working will be successful; nevertheless, by tracing current contradictions, experimenting, creating and exploring new possibilities of objectification, the approach may one day bear fruit in the effort to envision—and make—a postcapitalist future.

The aim of a revolutionary mathematics is to denaturalize statistics by demonstrating the social, economic and political stakes that are encoded in its functioning—in short, to intervene in metaphysical determinations. But this is not to deobjectify (or dereify) our situation, to end false consciousness by showing what "is really going on"—a task that appears to work fine for critical discourse but has a way of getting lost as soon as we find ourselves returning to the market or our computers. Statistics and machine learning can be put to other ends. Another objectification is possible; the ultimate task of a revolutionary mathematics is to seek this out.

This book is, in essence, about algorithmic society and informational capitalism. But in attempting to touch the metaphysical heart of contemporary global production, it also wrestles with philosophical and historical questions about the production of knowledge and the statistical and probabilistic forms that support that production. The first part traces the epistemological shifts that a data-driven economy have wrought. The second part provides a historical and philosophical overview of frequentism, the form of statistical inference that reigned throughout most of the twentieth century. The final part traces the Bayesian revolution in statistical and computational methods, along with the contradictions and possible futures it opens.

This short text is far from enough to revolutionize the way machine learning, the algorithmic production of knowledge and statistics function under late capitalism. However, I am hopeful that it can, at the very least, open for critique a series of problems that plague capitalism and data-driven knowledge production, and hint at the path future work could take toward a revolution in the production of both knowledge and value.

PART I
Ghosts of Departed Quantities

So may be Henry was a human being.
Let's investigate that.
. . . We did; okay.
He is a human American man.
That's true. My lass is braking.
My brass is aching. Come and diminish me, and map my way.

—John Berryman, "Dream Song 13"

Chapter 1
Automating Knowledge

In 2004, with Hurricane Frances charging toward the coast of Florida, Walmart's chief information officer, Linda Dillman, asked her team of data scientists to forecast the effects the hurricane would have on sales. Based on prior data, the team came up with a number of insights. In addition to predictable consumption increases like canned food, water and flashlights, they discovered that strawberry Pop-Tarts sales increase to seven times their normal rate immediately prior to a hurricane.[1]

The data Walmart used in its model was derived from Hurricane Charley, a storm that had struck the peninsula only a few weeks prior. Using this as a test case, Dillman's team was able to contrast consumption patterns between stores in the path of the hurricane to stores not in the path of the hurricane. And indeed, sales of many of the items they stocked prior to Frances based on these insights sold as expected.[2]

While Walmart's knowledge proved profitable in the near term,

---

1 Constance L. Hayes, "What Walmart Knows about Customers' Habits," *New York Times*, November 14, 2004.
2 Ibid.

its statistical correlations offer little insight into underlying causal mechanisms. And, even if Dillman might claim otherwise, such correlations offer us little meaningful understanding of the world. Walmart's model, built from data and algorithmically produced patterns discovered in that data, centers knowledge production on correlations. Their epistemic relationship to causality is irrelevant: the model cares little about *why* Walmart consumers prefer strawberry, so long as a mathematically modelable relationship exists.[3] Ultimately, what is produced by Walmart is particular, local only to the conditions of the given data, and therefore unable to express any enduring principles. The irrelevance of Walmart's model beyond a very specific and immediate problem demonstrates a certain limit of algorithmic knowledge, even if it is one with which engineers and researchers are rather comfortable.

All the same, these limits do not undermine the importance or value of machine learning models, because they are often designed to update themselves over time, offering up-to-the-minute predictions as new data comes in—a procedural analog to the realities of a dynamic, capricious society of capitalist consumption. With this self-correcting design, the algorithm generates not some ideal, universal understanding, but instead a local and constantly shifting set of predictions.

While Dillman and her team may have played some role in setting up the model and selecting the data, when it comes to the knowledge produced, their role is that of an envoy: they know only what a statistical model tells them. And that model can only suggest, with a given probability, that consumers will likely buy more strawberry Pop-Tarts prior to a specific hurricane. Due to its complexity, it is the computer-run model that in the end understands, rather than the researchers. It is in this sense a nearly automatic production of knowledge; even if humans are involved, they

---

3 Christian Sandvig, "You Are a Political Junkie and Felon Who Loves Blenders: Recovering Motives from Machine Learning," paper presented to the symposium, "Data Associations in Law and Policy," Faculty of Law, University of New South Wales, Sydney, Australia (December 11, 2015).

appear not to add intellectual labor to the production. Machine learning operates by taking a large set of inputs and computationally determining a way to map those inputs to outputs. Machine learning techniques most often work by running through a set of training examples, evaluating the output, and updating themselves in order to best match the training data.

We witness here, in miniature, a larger turn toward statistics, and probability in particular, in the management of contemporary capitalism. The use of probability to predict the most likely set of outcomes has become central to everything from logistics to advertisements to the stock market and beyond. But, probabilistic analyses are notoriously bad at dealing with systemic change. Frank Knight, one of the founders of the mid-twentieth-century neoclassical Chicago school of economics, used the terms "risk" and "uncertainty" to bring to light one critical challenge to this type of probabilistic knowledge.[4]

For Knight, "risk" describes a set of knowable probabilities that can be managed—for instance, the probability that someone will win the lottery, the survival rate for various diseases, or the chance that a drug will produce a certain side effect. Correspondingly, "uncertainty" describes elements whose probability one does not know—or those one chooses not to include in their calculations. For example, when calculating the odds that a single ticket will win the lottery, the equation normally does not include the probability that the state or organization running the lottery will go bankrupt. While risk can be calculated, uncertainty cannot. Uncertainty threatens every model, because there are always dangers that lie outside the closed space of the system. While historically, capitalism has excelled at integrating newfound uncertainties into its social and economic processes, the powers of machine learning threaten to upset this stability. For the more efficiently systems manage risk, the harder it becomes to imagine and prepare for uncertainty. By definition, efficiency comes at the expense of redundancies and the type of double-checking that could make it easier to handle unexpected situations. For an example of this, one

---

4 Frank H. Knight, *Risk, Uncertainty, and Profit* (Boston: Houghton Mifflin, 1921).

need look no further than capitalism's privileging of short-term profits blocking any real response to climate change.

Since at least 1876, statisticians have been aware of the effects, in calculations of probability, of their choices about what data to include. John Venn, the man whose intersecting diagram made his work famous to grade-schoolers, describes the origin of this problem: "Every individual thing or event has an indefinite number of properties or attributes observable in it, and might therefore be considered as belonging to an indefinite number of different classes of things."[5] This problem is now known as the reference class problem, and describes the challenges of how this "indefinite number" is narrowed down, and thus defined, into something more manageable. It is at its base an ontological question about what something is and what else falls into that category.

The reference class problem directly challenges the supposed objectivity of statistics and machine learning. For example, if a patient is diagnosed with cancer, how that patient is defined—such as what stage their cancer is—determines their calculated probability of survival. Just as there is no "true" patient demographic, there is no "correct" way to define the world, only decisions that frame that world in potentially disparate ways. While some approaches, especially Bayesian ones, claim to avoid the reference class problem, they too choose what data is relevant to a given problem.[6] Whoever defines the reference class—or selects the relevant data that is included in an analysis—wields extraordinary power over the seemingly "data-driven" neutrality of statistics. A statistical model like Facebook's News Feed algorithm, for instance, weds users to the variables Facebook employed to construct that model. The conditions of possibilities are prefigured according to Facebook. In this way, probability is always political.

Statistics can only operate within the closed world of the

---

5 John Venn, *The Logic of Chance* (New York: Macmillan & Co., 1866), 176.

6 Alan Hájek, "The Reference Class Problem Is Your Problem Too," *Synthese* 156, No. 3 (2007): 563–585.

reference class or the data it is given to operate on. Correspondingly, as an enclosed world develops and new statistical models describe that world with increasing accuracy, it becomes harder to imagine that anything might exist outside it and upset it. The greater one is able to manage risk, the more unlikely and unimportant uncertainty appears to be.

One of the great challenges to machine learning is that often, a wide variety of factors can explain any measured difference. In the Walmart model, for instance, geography, the time of week, or even the internal temperature of each store may have an effect. All algorithmic systems of knowledge production are relative only to the data that is put into the system. This was precisely what doomed Google's attempt to predict flu activity based on search results: the model worked well for the first few years, but in 2011, something likely changed in how people searched for flu information, because the model predicted a number of doctor visits for flu that was more than double the figure reported by the Centers for Disease Control and Prevention.[7] Again, the problem here was that algorithmic models are only able to assess risk (the probability that something will happen) relative to the input data, while remaining perennially vulnerable to uncertainty (the inability to know everything).[8] While this is, to an extent, true of all knowledge, the use of theories and the discovery of causal mechanisms help buttress our trust in predictions and extrapolate beyond the data at hand; in Walmart's model, the fact that the correlation discovered applies only to prior data means there is even less guarantee that some unforeseen uncertainty will not intervene to disrupt its predictive power.

While statistical modeling has been used since at least the 1700s

---

[7] Declan Butler, "When Google Got Flu Wrong," *Nature* 494, No. 7436 (2013): 155–156.

[8] Machine learning and its relationship to capitalism directly open the question of the "event" as a syncope in historical time that could change faster than an algorithm is able to respond or update. See Alain Badiou, *Being and Event*, trans. Oliver Feltham (New York: Continuum, 2006); and Jacques Derrida, *Writing and Difference* (London: Routledge, 2001).

to facilitate mathematical descriptions of relationships between phenomena for purposes of prediction, machine learning goes further, allowing for the modeling of significantly more complex relationships based on vast fields of data. By design, it also largely abandons the hope that one could extract some universal understanding from observed relationships. Traditionally, statistical models were built on relatively simple formulas for relating input variables to output variables. For example, in 2004 it was reported, based on data from the United States and United Kingdom, that on average, each additional inch of a person's height correlates with a $789 increase in annual salary.[9] While the study does not tell us why this happens, we nonetheless learn something about the relationship between height and income. The relationship is a simple, linear one. Given only two individuals' height, the expected income difference between both can be easily calculated with pen and paper.

In contrast to this ease, machine learning models generally rely on massive numbers of calculations—so large it would be impossible for a human to do them in a reasonable amount of time—in order to train an algorithm. Furthermore, these algorithms frequently account for nonlinear relationships (wherein a change in one variable does not correspond to a constant change in another) between an immense sea of variables. For instance, the predicted strength of a hurricane could significantly relate to Pop-Tart sales, but the amount of those sales drastically increases between Category 2 and 3 hurricanes, levels off at 4, and then declines prior to a Category 5 storm.

With enough data, machine learning algorithms are able to detect incredibly elaborate interactions between variables. But these interactions often become so complex that while the algorithm may find strong correlations and thus "learn" them, humans, for all intents and purposes, never can.[10] Unlike scientific theories

---

9 Timothy A. Judge and Daniel M. Cable, "The Effect of Physical Height on Workplace Success and Income: Preliminary Test of a Theoretical Model," *Journal of Applied Psychology* 89, No. 3 (2004).

10 Will Knight, "The Dark Secret at the Heart of AI," *MIT Technology Review*, April 11, 2017.

that offer elegant mathematical descriptions of universally valid causal processes, we ask machine learning systems something much more epistemically complex: to calculate the probability of a set of outcomes over a circumscribed field of examples. In response, those systems generate a set of probabilistic knowledges from which one can, in the aggregate, act. To allude to the epigraph from John Berryman, we initially determine probabilistically that Henry is a human being, then, with more data, that he is a human American man. With that, we may claim to begin to map his way.

Because these models are only as effective as the data input into them, the solutions and models that have been developed in the past decade tend to work well only in the relatively limited domains for which they are intended. Machine learning algorithms have effectively classified images, predicted purchasing behavior, turned audio into text, and even allowed digital assistants (such as Amazon's Echo) to provide an increasing array of voice-activated functionality. But the utopia (or perhaps dystopia) of an artificial general intelligence that can carry out abstract and complex general tasks remains far afield.[11] Moreover, the data on which these models are trained—and the biased social reality that the data represents—build into their algorithmic outputs simple replications of extant systems.[12]

---

11 "Artificial general intelligence" (AGI) refers to the capacity of machines to fully approximate or even surpass the capacity of human intelligence to perform complex tasks. Cassio Pennachin and Ben Goertzel define genuine AGI as a "software program that can solve a variety of complex problems in a variety of different domains, and that controls itself autonomously, with its own thoughts, worries, feelings, strengths, weaknesses and predispositions." "Contemporary Approaches to Artificial General Intelligence," in *Artificial General Intelligence*, Ben Goertzel and Cassio Pennachin, eds. (New York: Springer, 2006), 1.

12 Google's senior vice president for search and head of AI, John Giannandrea, has claimed that AI and machine learning's greatest threat to humanity are the biases machines and algorithms learn through training with already-prejudiced input data. Will Knight, "Forget Killer Robots—Bias Is the Real AI Danger," *MIT Technology Review*, October 3, 2017. On the proliferation of algorithmic bias, see Will Knight, "Biased Algorithms Are Everywhere, and No One Seems to Care," *MIT Technology Review*, July 12, 2017.

Because all data is mediated through our social world, biases cannot but saturate the closed worlds of algorithmic knowledge production. Consider, for example, the COMPAS (Correctional Offender Management Profiling for Alternative Sanctions) algorithm, designed by software company Northpointe, whose results are regarded by courts and prison systems as predictive of an individual's likelihood of recidivism. In 2016, investigative journalists from ProPublica determined that the algorithm was demonstrably racist. Based on a series of questionnaires and demographic data, COMPAS identified Black defendants to be at a significantly higher risk for reoffending while identifying white defendants at a much lower risk.[13] While the above Pop-Tarts example challenges how we understand knowledge in the abstract, COMPAS's racism makes the challenges of algorithmic knowledge much more concrete: algorithmic bias in predicting recidivism rates can have a significant impact on pretrial, parole, and sentencing decisions.[14]

The algorithmic logics that both Walmart and Northpointe use represent a new epistemic version of the world, one that is private and particular. And Walmart is far from the only company attempting to predict their customers behaviors: Target famously predicts when customers are pregnant;[15] Canadian Tire discovered that individuals who bought felt pads to protect their hardwood floors never missed credit card payments—while those who bought chrome skull accessories for their car were almost guaranteed to default.[16] Each of these companies builds their own private databases of sales or arrests (and buys databases from other vendors) to create microcosms of the world where they can continually compute the probability of various events.

---

13 Julia Angwen and Jeff Larson, "Bias in Criminal Risk Scores is Mathematically Inevitable, Researchers Say," *ProPublica*, December 30, 2016.

14 Julia Dressel and Hany Farid, "The Accuracy, Fairness, and Limits of Predicting Recidivism," *Science Advances* 4, No. 1 (2018): 1–5.

15 Kashmir Hill, "How Target Figured Out a Teen Girl Was Pregnant before Her Father Did," *Forbes*, February 2, 2012.

16 Charles Duhigg, "What Does Your Credit-Card Company Know about You?," *New York Times*, May 12, 2009.

While techno-futurists have sold a world of untold wonder in which computers and machines will predict our every need, both collectively and individually, the reality is much more constrained and disappointing. This is, in part, because many of these systems excel at interpolation (the process of filling in holes in given data), while progress on extrapolation (the prediction of future, completely unknown situations) has been limited.[17] We will explore the social implications of these technologies soon enough, but in order to better grasp what is at stake, it is worth explicating, first, what machine learning is and how it works.

Artificial Neural Nets

Many machine learning systems are built using the powerful, yet relatively straightforward *artificial neural network*. An ANN consists of multiple layers of artificial "neurons" that are connected to each other, in a structure that mimics a very simplified model of the human brain. In biology, at the distinct risk of oversimplifying things, human beings "learn" when synapses connecting their networked neurons fire between each other, increasing the ease with which similar signals can travel, and thus remembering, those pathways. In machine learning, the basic idea is similar: training data is fed through a network, and the network attempts to discover the best way to transform input values into outputs. ANNs are made up of multiple layers, where each neuron in a layer connects to each neuron in the following layer. A simple ANN can have an input layer, a single "hidden" layer and an output layer; more complex networks can have multiple hidden layers and complex divisions inside layers. In either case, each neuron tries to learn how to evaluate the information coming from its input neurons. In this way, every neuron—and the network as a whole—remembers

---

17 Georg Martius and Christoph H. Lampert. "Extrapolation and Learning Equations," *arXiv*, 2016.

its history in abstracted form, maintaining a trace of the values of what it has seen before, in order to create a model that can make predictions based on new data.

Imagine we create our own Pop-Tarts model, using data about the probability of a hurricane, maximum predicted wind speed and predicted rainfall. With this data, like Walmart, we might want to predict the increase in Pop-Tart sales through our model. Given that our input data includes probability of a hurricane, wind speed and rainfall, we compose our input layer with these three "neurons." And because we want to predict only the number of Pop-Tarts sold, we include a single output neuron representing that value.

Between the input and output layers is the "hidden" layer that allows the model to describe and predict complex relationships within the data. In our hidden layer, we can choose to include a certain number of neurons (see the figure below).

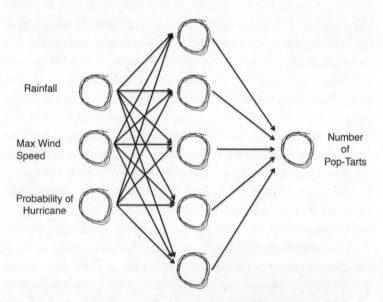

Diagram of a theoretical ANN to determine the relationship between rainfall, maximum wind speed and the probability of a hurricane and the number of boxes of Pop-Tarts sold.

Our five hidden neurons allow the ANN to detect different elements of our prediction problem. For instance, one of the five neurons could be on the lookout for hurricanes so devastating that inhabitants leave town rather than stock up on food. Thus, we would want that neuron to detect when all three inputs (probability of a hurricane, wind speed, and rainfall) are very high, and then suggest to the final output neuron to decrease the predicted number of Pop-Tarts sold. It would be rare, in practice, to have a neuron that so clearly detects something easily describable in human language; indeed, such neurons detect patterns in the input data that are normally more complex, dynamic and of a finer granularity than can be captured by human description. Nevertheless, this simplified example serves to illustrate how the composite set of neurons that constitute the system learn how to detect various features in the data and change the model's prediction based on what they observe.

The choice here of five neurons is largely arbitrary, although the addition of more neurons—and more hidden layers—generally conditions the ANN to be more sensitive to smaller, less perceptible patterns, at the cost of requiring more time to run and more data to train. Furthermore, the use of more hidden neurons and layers risks the problem of "overfitting," to which we will turn later.

Located within each hidden neuron is what machine learning researchers call an "activation function." In machine learning, this activation function is what defines the set of conditions under which that neuron "fires," mirroring, in some ways, how animal neurons activate. Most often, various nonlinear functions are used such that the activation function resists "firing" if it receives only a small stimulus. Then, at a certain measurable level, it rapidly increases its output signal and finally plateaus. This allows the network to detect nonlinear relationships, such as we see in our "extreme hurricane detector": initially, the neuron would add little information to the final prediction, but were an input to cross a certain threshold (say, a 90 percent chance of a Category 5 hurricane), the stimulus would quickly shift and activate the neuron, decreasing the predicted number of Pop-Tart sales.

Many things in the world display similar nonlinear dynamics. The rate at which people fall in love, income distributions, the popularity of works of art—all largely eschew the boring simplicity of linear relations (as X increases, Y increases at a steady rate). And, as these nonlinear elements interact with each other in nonlinear ways, the overall system becomes incredibly complex. Accordingly, when machine learning researchers attempt to predict real-world outcomes, the ability to account for such nonlinear dynamics becomes crucial.

In order to turn the network into a functioning model, we first assign arbitrary weights to each neuron's input. We then run a set of training data—data that already has the correct answer associated with it—noting how far away the model is on each prediction. After the training dataset is run, we average how far off the final neuron is from the correct answer that supervises the model. We then adjust the initial weights in order to minimize error, that is, the distance to the correct answer.

In this process of tweaking, we work our way backward through the network, revising the weights at each step in order to decrease the amount of error, in a process known as "backpropagation." After running data and backpropagating, the computer does it again. And again. And again, working through all of the training data over and over. Machine learning algorithms repeat this process thousands of times, slowly adjusting these weights until subsequent rounds of training find stability. We cannot simply calculate the optimal weights, since the ANN network is strongly interconnected, and changing a single neuron's weight alters the optimal value for all other weights.[18] When those backpropagated

---

18 As a result of the complexity of many of these algorithms, they oftentimes succeed in finding locally optimal solutions to problems, which means that under different starting conditions but with the same data, a different solution could be reached.

weights finally stop moving, the model has converged: in short, it has come up with a solution to the training data and can take new data to give us an algorithmically produced "answer." This process of backpropagation is extremely important, both computationally and historically; as computer scientist Ethem Alpaydın has noted, while the basic idea of computational neural nets existed for most of the second half of the twentieth century, it was not possible to train multilayer ANNs until the invention and further development of backpropagation throughout the '70s and '80s.[19]

The amount we adjust each neuron's weight during backpropagation is known as the *learning rate*. The higher the learning rate, the more quickly the network converges. In the case of large networks that use large training datasets, the time required to train a model can be considerable, ranging from a couple hours to several days.[20] If we quicken the learning rate, we tend to find less accurate results, with the weights swinging wildly in each round. Our overeagerness, or what is referred to as *overgeneralization*, might quickly lead the model, upon seeing only a few Category 5 hurricanes in the dataset, to decide that all hurricanes are Category 5. In this way, we risk becoming guilty of Hegel's famous critique of Schelling's philosophy—that is, of simply learning that at night, "all cows are black."[21]

Opposite overgeneralization is a related technical challenge for machine learning: the problems of *overfitting*. If a model is overfitted, we have made it too smart, merely memorizing the training data, much like a student who learned only how to take the practice test. This rigidity results in a model that may be very effective at correctly analyzing the training dataset, but remains stubbornly unable to employ that information outside of the confines of the input model.

---

[19] Ethem Alpaydın, *Machine Learning: The New AI* (Cambridge, MA: MIT Press, 2016), 99.

[20] Josh Patterson and Adam Gibson, *Deep Learning: A Practitioner's Approach* (Sebastopol, CA: O'Reilly Media, 2017), 321.

[21] Georg Wilhelm Friedrich Hegel, *The Phenomenology of Spirit*, trans. A.V. Miller (Oxford: Oxford University Press, 1977), 9.

In either case, the models' computationally produced abstractions fail to generalize what they have learned effectively enough to predict the correct answer when confronted with new, not-on-the-test data. In order to generate useful predictions, machine learning must walk the fine, functionalist line between generalizing too much and not enough.[22] This situation, along with the contemporary moment's growing availability of computing power, more extant algorithms to choose from and more stores of data, means that we encounter a dual possibility: better predictions, on one hand, and an increased danger of overfitting or overgeneralization, on the other.

The dangers of overfitting and overgeneralization bear a formal similarity to the reference class problem described earlier: How to make a set of meta-level decisions in order to generate a model that best describes the world? But, unlike the reference class problem—where one can have sensible debates about what type of thing something is for the sake of calculation—the decisions inherent in machine learning practice are significantly more opaque, because small changes can have outsized effects and we often do not know what elements in a dataset an algorithm has decided are relevant.

These decisions, like the learning rate or the number of neurons and hidden layers, are generally set by humans during the construction of the model rather than learned as part of the process. Known as "hyperparameters," their human-selected fine-tuning can make authoring machine learning algorithms more of an art than a science. And as machine-learned systems model more and more complex nonlinear dynamics, the impact of changes to hyperparameters are, consequently, also complex,

---

22 This focus on the utility of prediction or the predictive capacity of theoretical models, particularly in spite of their often overly simplistic assumptions about the nature of the world, is a scientific vision of the world that is also embraced by financial theorists like Milton Friedman. See MacKenzie, *An Engine Not a Camera*.

nonlinear and largely unpredictable.[23] Increasingly, this means that the outputs of these systems are not easily verifiable or testable.

At its most basic, our hurricane-detecting ANN takes a high-dimensional input space (only three in this example, but often many more in practice) and maps the inputs through a series of learned nonlinear transformations to a lower-dimensional output space (our single "Pop-Tart prediction").[24] Here we cross one of the central taxonomic distinctions in machine learning: that between supervised and unsupervised learning. *Supervised learning* takes place in situations where training data allows the algorithm to learn from examples where the answer is known, such as in the case of the training data for our hurricane ANN. *Unsupervised learning*, on the other hand, involves a search for underlying structures in data where no "correct" answers are predefined. Classification algorithms, which seek to divide data into a number of categories, are a common example of this type of learning.[25]

While this distinction is significant in practice, the most important point for our present concerns is that both take large, though still human-selected, datasets and try to find structures

---

23 Google has recently succeeded in making a machine learning algorithm to optimize hyperparameters for its "child" algorithm, which have outperformed the best human-tuned models by a little bit. But still, the original algorithm requires well-tuned hyperparameters. Dom Galeon and Kristin Houser, "Google's AI Built Another AI That Outperforms Any Made by Humans," *Futurism*, December 1, 2017.

24 In general, machine learning problems come in two distinct types. The first are regression problems, such as the one we have been describing here, which involve predicting some numerical value. The second are classification problems, where we want to know if something is of one type or another (e.g., whether a sports team will win or lose). The general approach is the same, but in this case, we are trying to determine if the output is zero or one, rather than somewhere on a spectrum, such as the price of Pop-Tarts.

25 One example of this type of algorithm is "k-nearest neighbors," which classifies data by finding points in the dataspace that can be used to efficiently group the data.

in the data in order to map internal similarities (in the case of unsupervised learning) or a known output (in the case of supervised learning). Despite the differences between these two methods, both supervised and unsupervised learning attempt to find some signal that exists in the data—in essence, to find something meaningful.

Though saturated in complex implementations and large datasets, this rapidly growing field—and the massive financial and human investments it has enticed—is built on these relatively simple conceptual foundations. Despite an often-convoluted, "experts only" discourse, machine learning requires no complex symbolic logic, no deep underlying theory of the mind, and no universal understanding of the nature of the world.

In fact, these machine-learned models apprehend a very narrow, particular vector of the world, based only on the observed data and the nonlinear relationships discovered within it. While traditional ANNs do not necessarily deal directly in probability, epistemically and metaphysically they partake of this general turn to a probabilistic understanding of the world and knowledge: these systems are not designed to be correct all the time, but rather to guess the correct outcome with a high enough probability that en masse, the system is economically viable.

"Indeed, They Don't Have to Settle for Models at All"

In many ways, machine learning was born out of disappointment: the failure of artificial intelligence to live up to its postwar promise. During the 1960s, many American attempts at AI sought to use symbol-based systems to solve logic problems, mimic human language, translate between languages, and the like. These systems attempted to understand classes of things logically and how different classes of things interact, such that conclusions could be worked out from initial premises. While early constrained attempts at these problems proved encouraging, attempts to take them beyond simple problems quickly proved intractable.[26] Early optimism, such as computer scientist Marvin Minsky's 1967 claim that "within a generation . . . the problem of creating 'artificial intelligence' will substantially be solved"[27] was soon quashed, and considerable government resources gave way to an "AI winter" where public interest, funding and research efforts stalled.[28]

Apprehending the limitations of these symbolic attempts, researchers turned to more probabilistic and statistical approaches, such as ANNs. The success of these approaches in subsequent decades had its roots in earlier postwar science. The history of neural networks themselves begins in 1943, when cyberneticians and neuroscientists, Warren McCulloch and Walter Pitts, published "A Logical Calculus of the Ideas Immanent in Nervous Activity," which proposed a representational model of human knowledge as the activity of networked neurons. According to McCulloch and Pitt's "neuron" model, these neurons functioned by summing together binarily represented input data, where knowledge was subsequently represented as a binary output, depending on whether or not the sum of inputs reached a certain threshold. This model profoundly, albeit simply, imagined the brain as a computational machine, where "any computable function could be computed by some network of connected neurons, and . . . all the logical connectives (and, or, not, etc.) could be implemented by simple net structures."[29]

In 1958, AI pioneer Frank Rosenblatt expressed disappointment at the simplicity of models like the one created by McCulloch

---

26 Bruce G. Buchanan, "A (Very) Brief History of Artificial Intelligence," *AI Magazine* 26, No. 4 (2006): 58–59.

27 Marvin Minsky, *Computation: Finite and Infinite Machines* (Englewood Cliffs, NJ: Prentice-Hall, 1967).

28 For an exceptional essay on how machine learning has changed and could change our understanding of intelligence, see Catherine Malabou, *Morphing Intelligence: From IQ Measurement to Artificial Brains* (New York: Columbia University Press, 2019).

29 Stuart Russell and Peter Norvig, *Artificial Intelligence: A Modern Approach*, 3rd ed. (Upper Saddle River, NJ: Pearson, 2010), 16.

and Pitts, which relied on Boolean algebra to represent neuronal activity. By representing input data as only a registered one or unregistered zero, these models, Rosenblatt alleged, failed to demonstrate how the human brain processed more complex information, and its ability to represent the world in gradations.

In response, Rosenblatt proposed his own "theory of statistical separability" to facilitate a more dynamic representation of a neural network.[30] In this dynamism, one could account for weighted input stimuli, where data was not just a one or zero but a probabilistic stimuli that could attend to small statistical differences.[31] Using this concept of statistical separability, Rosenblatt developed a network he called the "perceptron," which was able to handle much more complex information than its predecessors and, precisely due to the probabilistic, nonbinary representation of its inputs, could potentially learn through feedback from trial-and-error tests.

As excitement around the symbolic approaches continued to grow, Rosenblatt's perceptron fell into disfavor. In 1969, computer scientists, Marvin Minsky and Seymour Papert, published their *Perceptrons: An Introduction to Computational Geometry*, a work that was largely critical of the statistical separability approach, demonstrating that it could not meaningfully solve even simple problems.[32] Their book hastened the abandonment of neuron-based attempts for the next decade.[33]

---

30 Frank Rosenblatt, "The Perceptron: A Probabilistic Model for Information Storage and Organization in the Brain," *Psychological Review* 65, No. 6 (1958): 386–408.

31 Ibid., 388.

32 In their book on perceptrons, Marvin Minsky and Seymour Papert question the extent to which a perceptron could effectively learn in a manner analogous to the human brain due to its incapacity to solve the "XOR," or "exclusive or," problem. *Perceptrons: An Introduction to Computational Geometry* (Cambridge, MA: MIT Press, 1969). It has since been proven that multilayer perceptrons can in fact solve this problem by including a hidden layer.

33 Pamela McCorduck, *Machines Who Think: A Personal Inquiry into the History and Prospects of Artificial Intelligence* (Natick, MA: AK Peters / CRC Press, 2009), 106–107.

But in the '80s, further developments in methods for working with neural networks, including backpropagation and the explosion of cheap computing power, provided the conditions for a return to probabilistic, neuron-based approaches to AI.[34] In addition to the rise of neural networks, other nonsymbolic approaches—from the "naive Bayes classifier" to support-vector machines—appeared, employing increasing computing power and methodological advances to generate knowledge and intelligence without relying on a fixed idea of meaning.[35] Rather than symbolically deducing conclusions from premises, as older symbolic approaches did, many of these newer systems were able to count evidence probabilistically, and thus provide a probabilistic output, ultimately favoring so-called "temporary correlation" over an understanding of causality or foundational principles.

With their movement away from the disappointing symbolic systems and toward statistics and probability, these newer machine learning methods reframed the epistemic contours of computationally produced knowledge, allowing its automation by abandoning its fixity. Rather than symbolically creating some universal, generalized representation of knowledge and the world, today's explosion of machine learning instead generates models that are allowed to develop their own, problem-specific internal logics. In a way, machine learning is witnessing the development of an array of near-infinite, hyper-specific enlightenments, of highly regulated simulations able to process the world—albeit without any connection to a notion of the universal—by creating them anew for each particular place and time. Our hurricane-detector ANN calculates one world, while someone else's, with only slightly different data or structure, calculates another.

---

34 David E. Rumelhart, Geoffrey E. Hinton and Ronald J. Williams, "Learning Representations by Back-propagating Errors," *Nature* 323 (1986): 533–536.

35 Ethem Alpaydın, *Introduction to Machine Learning*, 3rd ed. (Cambridge, MA: MIT Press, 2014).

In a short article in *Wired*, Chris Anderson, a Silicon Valley evangelist made rich by his undying faith in machine learning technology, spells out the theoretical implications of this development. "Today companies like Google, which have grown up in an era of massively abundant data, don't have to settle for wrong models," he writes. "Indeed, they don't have to settle for models at all."[36] He continues, repeating a realization that Hegel noted long before modern computation: at a certain point a change in quantity becomes a change in quality.

> The Petabyte Age is different because more is different... At the petabyte scale, information is not a matter of simple three- and four-dimensional taxonomy and order but of dimensionally agnostic statistics. It calls for an entirely different approach, one that requires us to lose the tether of data as something that can be visualized in its totality. It forces us to view data mathematically first and establish a context for it later. For instance, Google conquered the advertising world with nothing more than applied mathematics. It didn't pretend to know anything about the culture and conventions of advertising—it just assumed that better data, with better analytical tools, would win the day. And Google was right. Google's founding philosophy is that we don't know why this page is better than that one: If the statistics of incoming links say it is, that's good enough.[37]

A "good enough" probabilistic world ousts those traditional Enlightenment models and theories of causality. While it is beyond doubt that the efficacy of machine learning, at least in certain applications, is impressive, these technologies and their ideological commitments are not as direct or unmediated as they are often

---

36 Chris Anderson, "The End of Theory: The Data Deluge Makes the Scientific Method Obsolete," *Wired*, June 23, 2008, 16–17.
37 Ibid.

made out to be.³⁸ What Anderson presents as a direct apperception of the real is, in fact, mediated through decisions about how to construct models, how to avoid overfitting and how to understand what those results tell about the world.³⁹

The possible advantage to this correlational approach, as informatics scholar Geoffrey Bowker argues, "is that it avoids funneling our findings through vapid stereotypes. Thus, in molecular biology, most scientists do not believe in the categories of ethnicity—and are content to assign genetic clusters to diseases without passing through ethnicity (e.g., Karposi's sarcoma as initially a Jewish disease)." Yet, while many social categories are founded on questionable premises—race, gender, value, and so on—the histories and impacts of these categories are still highly important because "the world is structured in such a way as to make the categories have real consequences."⁴⁰ Although these categories are problematic, to do away with them completely would leave us unable to identify these real consequences.

Media theorist Wendy Hui Kyong Chun's recent work on the concept of homophily—the love of the same—is especially insightful in this regard.⁴¹ Focusing on the use of network science as an analytical and computational technique, she argues that these

---

38 Davide Panagia agues that algorithms ultimately manage variability and probability in order to dispose relations, such that "these dispositional powers are operant regardless of any epistemic reform one might adapt to its computational logics. In short, an algorithm is a dispositif not because it constrains freedom through various forms of domination, but because it proliferates controls on variability and, in this way, governs the movement of bodies and energies in space and time." "On the Possibilities of a Political Theory of Algorithms," *Political Theory* 49, No. 1 (2021): 109–133.

39 It is worth noting, in this regard, that Andrew Gelman, one of the leading Bayesian theorists of our time, constantly describes the process of inference as one of model building and testing. See, for instance, "Bayes, Jeffreys, Prior Distributions and the Philosophy of Statistics," *Statistical Science* 24, No. 2 (2009): 176–178.

40 Geoffrey Bowker, "Big Data, Big Questions: The Theory/Data Thing," *International Journal of Communication* 8 (2014): 5.

41 Chun, "Queerying Homophily."

systems push people toward those who are the same, creating digital spaces that are segregated along the multiple and intersecting vectors of social existence. And while these digital systems are more mobile and fluid than earlier forms, she states, "there are no random initial conditions."[42] These digital systems are built on histories and in societies that have long and complicated histories of racial and gendered violence. Thus, it was "the rise of the modern concept of race during the era of Enlightenment; its centrality to colonization and slavery; its seeming zenith during the era of eugenics; [and] its transformations after World War II" and beyond that set the initial conditions for these systems and determine the social world they tend to reproduce.[43]

Despite what Anderson claims, mathematics routed through machine learning is still, in the end, a form of mediated abstraction, and thus fundamentally intertwined with the development of our social, economic and political situation. These machine learning systems function in a way that is analogous—and, as we shall see, metaphysically tied—to capitalism: they move the locus of social domination from the material world into the abstract one of capital and probability, yet they do not oust history. In fact, because their aim is only to predict, they actively reproduce it. While these systems make some categories more fluid and open, they simultaneously work to solidify extant social systems: capitalism, racism, patriarchy and imperialism, among others. And they do so in ways that are potentially more insidious and harder to resist, presenting their outputs as objective facts. It is thus necessary to account for the metaphysical force of this objectification—something we can ascertain only by tracing the ways in which probability and statistics function socially and economically.

---

42 Ibid., 82.
43 Ibid., 84.

Chapter 2
Can Computers Do Math?

From the sheer number of algorithms affecting our daily life, to the amount of data that machine learning algorithms process, to the number of hidden layers within a model and the speed at which they run, the complexity of these nonlinear, machine-learned models prevents human beings from fully understanding what is happening inside them. What's more, the internal logic and state of these systems are only becoming less accessible.

The inaccessibility of algorithms is often described in terms of "black boxes," a characterization that tends to elicit political attempts to shine the light of transparency on algorithmic systems and slow their functioning to a speed that can be managed.[1] Yet, both algorithms themselves and the larger social systems within which they function continually resist these calls for transparency, in large part because of the speed with which they operate

---

1 Frank Pasquale, *The Black Box Society: The Secret Algorithms That Control Money and Information* (Cambridge, MA: Harvard University Press, 2015); Christian Sandvig, Kevin Hamilton, Karrie Karahalios and Cedric Langbort, "Auditing Algorithms: Research Methods for Detecting Discrimination on Internet Platforms," *Data and Discrimination: Converting Critical Concerns into Productive Inquiry* (2014): 1–23.

and are created. While the Volkswagen defeat device discussed in the introduction was finally regulated, it took years before its existence was even discovered. For each algorithm that is regulated or draws public ire, many more likely go unnoticed. If algorithms are designed largely to maximize profit or automate an already-biased system, the political issue must be both the algorithm and its larger context. As seen in sexist and racist outputs calculated from avowedly "neutral" algorithms, the unjust epistemic terrain of daily life tends to get baked into each machine-learned system and model, as well as the data from which algorithmic insights are drawn. Often, algorithms end up simply automating the bureaucratic opacity and injustice they are meant to replace. Thus, even attempts to decode how these systems make their decisions are unlikely to overturn the unjust social systems they operationalize.

Political projects that critically assess how algorithms produce and reproduce these systemic biases are critical, and organizing around these issues is decidedly important. However, these projects of algorithmic dereification are unlikely to address the unique political and economic challenges of our contemporary moment. To dereify algorithms, or to establish a politics around the revelation of what algorithms are "really doing," would ultimately require a capacity for regulation and critique that could overcome the ways in which capitalism poses and solves social problems. Calls for such regulation assume that were humans only to know what is happening in these systems, they would surely correct them; in short, they underestimate the metaphysical work upon which these systems are built, the power of exchange and capital that make their work appear necessary.[2]

---

2 Slavoj Žižek, citing Peter Sloterdijk's definition of cynicism, states that the cynical logic of modern capitalism can be characterized thus: "They know very well what they are doing, but still they are doing it." *The Sublime Object of Ideology* (London: Verso, 1989), 32–33. See also Mark Fisher, *Capitalist Realism: Is There No Alternative?* (London: Zero Books, 2009).

A politics of transparency may help prevent the worst abuses of algorithmic knowledge and is thus a worthwhile political project, but transparency alone can never completely solve the injustices algorithms present. As we saw with the discovery and censure of Volkswagen's defeat devices, there remains considerable benefit to regulation, yet such behavior cannot be regulated if it is able to escape detection in the first place. As long as there is a strong economic incentive to use algorithms for obfuscation, capitalist profit-seeking will continue to produce them. In sum, algorithms change the speed and form of knowledge production, but they do not change the ultimate goals of knowledge production.

In the complex ecologies in which algorithmic systems function, all sorts of economic, political and social incentives shape how they work and what they value. What forms the opacity of an algorithm, or knowledge production in general, is simultaneously that which is hidden inside the system, and those obfuscated and objectified forces that work from without to shape the necessities to which knowledge production responds. From the intricacies of contemporary mathematical proofs to the predictive power of neural networks, this means that all knowledge production is in essence "black boxed" because it relies on abstractions whose nature must be ignored in order to be useful, especially at the moment it becomes economically productive. It is precisely this withdrawal into unseen work that gives these systems their very force, the force of objectification.

Thus, in order to engage algorithmic culture politically, it is imperative to understand its abstract force. The question at hand concerns not only *how* these algorithmic systems function but *what* metaphysical force they produce that gives their outputs such authority in contexts ranging from courtrooms to boardrooms to bedrooms. To begin to ask these questions, however, requires a slight detour through the realm of theoretical mathematics.

Four-Color Theorem

Even theoretical mathematics encounters this problem of opacity that contaminates math's supposedly unadulterated internal logic. For example, in 1976 Kenneth Appel and Wolfgang Haken announced that they proved the famous "four-color theorem" (4CT). This theorem, proposed over a hundred years prior, states its problematic simply: any map divided into areas can be colored using only four unique colors without the use of a single color for two adjoining countries (provided that certain constraints are followed in the construction of the map such as disallowing discontiguous areas).

Historically, no one had found or fabricated such a map requiring more than four colors. But even with this lack of empirical evidence, it had never been mathematically proven that it was impossible. The theoretical possibility of a necessarily five-color map remained, and this possibility frustrated mathematicians for over a century.

Appel and Haken's proof took a novel approach: they used a computer. They were able to prove that all possible maps could be reduced to versions of a core set of 1,936 maps. If those maps could be colored with four colors, they argued, then so could any map. They then programmed a computer to check these nearly 2,000 maps to determine if they could be colored with only four colors—a task that took their 1970s machine over a thousand hours to complete.[3]

Following Appel and Haken's success, a host of other proofs have been computationally resolved. In 2016, three computer scientists produced a computer-assisted proof to the so-called "Boolean Pythagorean triples problem" whose final, written-down form consisted of 200 terabytes of data.[4] The resulting proof took

---

[3] Kenneth I. Appel and Wolfgang Haken, *Every Planar Map Is Four Colorable* (Providence, RI: American Mathematical Society, 1989); Robin J Wilson, *Four Colors Suffice: How the Map Problem Was Solved* (Princeton, NJ: Princeton University Press, 2002).

[4] Marjin J.H. Heule, Oliver Kullmann and Victor W. Marek, "Solving and Verifying the Boolean Pythagorean Triples Problem via Cube-and-Conquer," in *Theory and Applications of Satisfiability Testing—SAT 2016*, Nadia Creignou and

Ghosts of Departed Quantities

up twenty times the digital space of the entire English-language Wikipedia.[5]

Much like the impossibility of confirming the veracity of every English Wikipedia page, the very fact of such an enormous amount of data employed for one specific problem precludes the possibility that a small team of human researchers could ever fully certify the proof. Instead, proofs like the Boolean Pythagorean triples problem are verified by another piece of software. This is a perfect example of the epistemics of opacity: a computationally produced proof is read, and verified, by another computer.[6] Such theoretical knowledge becomes fundamentally recursive—it is verified simply by being repeated in a computer elsewhere. While algorithms for solving proofs are unique to the problems of pure mathematics, and thus differ substantially from common machine learning algorithms, the epistemic challenges they raise remain indicative of the larger state of computationally produced knowledge.

Indeed, not long after Appel and Haken proved the 4CT, mathematicians began questioning the value of such computationally facilitated endeavors. For instance, Thomas Tymoczko, a philosopher of mathematics, wrote an article the same year Appel and Haken's proof was published, in which he suggested that for such a "proof" to truly count, it would require a radical change in mathematicians' core definition of proof itself:

> No mathematician has seen a proof of the 4CT, nor has any seen a proof that it has a proof. Moreover, it is very unlikely that any mathematician will ever see a proof of the 4CT.
>
> What reason is there, then, to accept the 4CT as proved? Mathematicians know that it has a proof according to the most rigorous standards of formal proof—a computer told them.

Daniel Le Berre, eds., *Lecture Notes in Computer Science*, Vol. 9710 (London: Springer, 2016), 228–245.

5 Evelyn Lamb, "Two-Hundred-Terabyte Maths Proof Is Largest Ever," *Nature*, May 26, 2016.

6 Ibid.

Modern high-speed computers were used to verify some crucial steps in an otherwise mathematically acceptable argument for the 4CT, and other computers were used to verify the work of the first.

Thus, the answer to whether the 4CT has been proved turns on an account of the role of computers in mathematics. Even the most natural account leads to serious philosophical problems. According to that account, such use of computers in mathematics, as in the 4CT, introduces empirical experiments into mathematics. Whether or not we choose to regard the 4CT as proved, we must admit that the current proof is no traditional proof, no a priori deduction of a statement from premises. It is a traditional proof with a lacuna, or gap, which is filled by the results of a well-thought-out experiment. This makes the 4CT the first mathematical proposition to be known a posteriori.[7]

In the radical redefinition of proof required to include 4CT, human knowledge of the proof's specifics is no longer required. Contrary to even the most elementary empiricism, while we cannot "see" the proof, nor "see" the proof of the proof, we know that the proof exists. Philosophically, the pristine realm of pure logic and the a priori has been besmirched by the "experimental," the observed and the a posteriori. We know the proof exists, but we have yet to comprehend it. In the race for knowledge, the computer—and its use of large-scale processing—has outpaced us.

As math has become experimental, the conditions of an experiment, or the computational work of a proof, have become vitally important in the derivation of a general law.[8] The specific computer that is used, the prevention of errors in the code, and so on all partake of the final result. Accordingly, as soon as the realm of the

---

[7] Thomas Tymoczko, "The Four-Color Problem and Its Philosophical Significance," *Journal of Philosophy* 76, No. 2 (1979): 57–83.

[8] See Louis Althusser, "On the Materialist Dialectic," in *For Marx*, trans. Ben Brewster (London and New York: Verso, 2005), 161–185.

pure, ideal a priori is punctured, we fall back into the world of materiality and economy, and their objective force. It is due to the fact that all of these elements—in short, the experiment—are commensurate with the produced knowledge that the final result appears to be objective—even if we are not party to their computation.

The ABC Conjecture

While the proof of the four-color theorem requires we believe a computer, we see a similar structure of opacity at work even in nonautomated mathematics. In 2012, a somewhat-reclusive mathematician by the name of Shinichi Mochizuki published four papers on his personal website. These papers totaled around 500 pages and were the culmination of years of largely independent work. Mochizuki made no announcement of their release, but a colleague of his at Kyoto University's Research Institute for Mathematical Sciences noticed their appearance and alerted other mathematicians in the field.

Among several other proofs, Mochizuki's fourth paper claimed to prove, for the first time, a significant theorem known as the "abc conjecture." This theorem refers to an intricate property of algebraic equations that take the form of $a + b = c$. In order to resolve this previously impenetrable conjecture, Mochizuki invented a whole new type of mathematics, which he named "inter-universal Teichmüller theory."[9]

This inter-universal Teichmüller theory is so abstract, and ventures so far away from traditional math, that other mathematicians—even those working in the same subdiscipline as Mochizuki—have been unable to fully verify the proof. Mochizuki has since estimated that it would take a graduate student ten years to learn and understand the theories he has created. Adding

---

9 Davide Castelvecchi, "The Biggest Mystery in Mathematics: Shinichi Mochizuki and the Impenetrable Proof," *Nature News*, October 7, 2015.

further complication, Mochizuki has largely refused to attend international conferences or invest much energy in explaining the work to other mathematicians. Even those who have made headway on this task find themselves at a loss to explain it to others. For example, one anonymous mathematician stated, "Everybody who I'm aware of who's come close to this stuff is quite reasonable, but afterwards they become incapable of communicating it."[10]

We might raise similar objections, as Tymoczko did to Appel and Haken's 4CT proof, about Mochizuki's work. Perhaps Mochizuki has "seen" the proof, alongside two or three other people who have worked through it. But have mathematicians as a field seen it? Given that the number of people who can understand "regular" cutting-edge mathematics is relatively small, what is the required quorum for truth? How many people need to comprehend a proof for it to be a proof in the classical sense that Tymoczko suggests?

We can inquire further: Does a computer-resolved proof verified by another computer count more, or less, than a likely but unverified proof by a well-regarded human expert? Regardless of how one answers this question—and it is unlikely there exists one correct answer—any answer ultimately relies on economics and exchange. While proofs, of course, are not directly bought and sold, their production requires networks of verifiability—other mathematicians, other computers and so on—that determine when the threshold of proof has been passed, that is, when a collection of experimental data is commensurate with a proof. While it is likely that pure a priori knowledge has always been contaminated by an exchange that makes separating the a priori and the experimental impossible, with the advent of modern computing, the size, speed and scope of these forms of unverifiable knowledge now reveal the centrality of exchange. It is because, not in spite, of these material networks of verification that this knowledge confronts us as objective, as this materiality allows it to appear to stand outside of our social world and account for itself.

---

10 Ibid.

The Infidel Mathematician

These problems of epistemology are certainly not new. Questions of how and what is required to "know" things have plagued humans since the earliest moments of philosophy. An especially telling example of this problem is British philosopher and bishop George Berkeley's 1734 essay decrying calculus entitled *The Analyst*, with its accompanying (rather long) subtitle: *A Discourse Addressed to an Infidel Mathematician: Wherein It Is Examined Whether the Object, Principles, and Inferences of the Modern Analysis Are More Distinctly Conceived, or More Evidently Deduced, than Religious Mysteries and Points of Faith*. Aside from its attempts to wrestle with the foundations of knowledge and attack calculus, Berkeley's *Analyst* and its questions also play an important role in the history of computational knowledge and capitalism.[11] The Reverend Thomas Bayes, for whom the field of Bayesian statistics is named, wrote only two texts during his life: one was a theological proof that God wants us to be happy, and the other a direct response to Berkeley's attack on calculus laid out in *The Analyst*. The following century, even Marx would take up the philosophical challenge of finding a solid ground for calculus, writing a short tract attempting to justify its mathematics.[12]

Berkeley's *Analyst* is addressed to an individual he calls the "infidel mathematician." Berkeley attacks this infidel's nascent calculus on multiple grounds. For one, he takes issue with the idea that one can, in modern parlance, take the derivative of the derivative. While one can describe what acceleration *is* (the derivative of velocity over time), Berkeley asks: What would the acceleration of an acceleration of an acceleration *mean*? Thus, he appears troubled by the same epistemic groundlessness that we confront when

---

11 George Berkeley, *De Motu and The Analyst: A Modern Edition with Introductions and Commentary*, trans. Douglas M. Jessup (New York: Springer, 1992).

12 Karl Marx, *Mathematical Manuscripts of Karl Marx*, Sofya Yanovskaya, ed. (London: New Park Publications, 1983).

thinking through the implications of recursive attributions of value and knowledge in algorithmic processing.

Berkeley further attacks the use of "infinitesimals" in this protocalculus—whose size approach zero when calculating the formulas for derivatives—as "ghosts of departed quantities." He argues that if they approach nothingness, they should provide no mathematical insight; still, mathematicians use these values as they approach zero to derive the basic formulas of calculus.[13] They function, in short, by making an *inequality equal*: by saying that absence is presence. While calculus has clearly done just fine for itself despite these attacks, there is something deeply important about Berkeley's recognition of the ways in which this knowledge requires the equalization of an inequality.

So much of modern knowledge is founded on such present absences, which require that what is represented is simultaneously the same and different—from diverse types of labor that are translated into value to the extraction of computable data from complex social interactions.[14] This is precisely what is at stake in

---

13 The following century, Augustin-Louis Cauchy formalized the concept of a limit and put the foundations of calculus on much more solid footing.

14 Similar mysteries can be found in the less sacerdotal branches of contemporary computation. For example, in 1985 the Institute of Electrical and Electronics Engineers (IEEE) approved "Standard 754" to regularize floating-point operations (i.e., operations including numbers with a decimal point). One of the more revealing elements in this standard is an insistence that NaN, which stands for "not a number," should not be equal to itself. A NaN is most often generated in response to an "illegal" operation, such as the division of a number by zero, so that a piece of software can gracefully deal with invalid instructions. The logic behind making NaN not equal to itself is to avoid accidental equivalence. If a program is supposed to compare two numbers, and both of them are the result of illegal operations, it might continue as though everything is okay until the machine crashes as more and more is done atop those defective operations. Perhaps the IEEE's Standard 754 is a rather sensible decision, and does not amount to a logic-defying mystery of faith. But we see here the importance of a standard that formally makes the equal unequal—and conversely the unequal equal—for the operable success of calculation. Dan Zuras, Mike Cowlishaw, Alex Aiken et al., "IEEE Standard for Floating-Point Arithmetic," *IEEE Std 754-2008* (2008): 1–70.

the redefinition of mathematical proof to include algorithmic proofs: the supposed presence of a formal proof that can be understood by a single mathematician is made commensurate with 200 terabytes of data that cannot be reviewed except by another computer. We will see these ghosts of departed quantities appear again and again as the fundamental unit of algorithmic knowledge production—a value that is both itself and its opposite (such as Bayesian analysis's claimed transubstantiation of the particular into the universal, or the subjective into the objective), able to make the incommensurate commensurate, and thus calculable.

To return to Berkeley, despite reservations about its philosophical foundations, Berkeley readily admits that calculus "works." Still, he argues that the faulty intellectual foundations of this math call into question its intellectual value. To the proponents of calculus, he says:

> Your Conclusions are not attained by just Reasoning from clear Principles; consequently, that the Employment of modern Analysts, however useful in mathematical Calculations, and Constructions, doth not habituate and qualify the Mind to apprehend clearly and infer justly; and consequently, that you have no right in Virtue of such Habits, to dictate out of your proper Sphere, beyond which your Judgment is to pass for no more than that of other Men.[15]

Even though this new math can be used to build bridges and solve otherwise-intractable problems, Berkeley does not agree with its fundamental premises. And, perhaps for him most importantly, he wants these mathematicians to focus on their math, rather than extrapolate from their insights there into the world of theology.

In essence, the operation Berkeley attempts to perform is one of dereification, stripping away the layers of this new math to show that its supposedly solid foundations are, in fact, insecure. But the

---

15 Berkeley, *The Analyst*, 209.

reason calculus is now taught throughout the world—while Berkeley's text is rarely read—is that calculus works and has worked, despite the supposed shakiness of its foundations. Likewise, capitalism appears to work with or without individual knowledge of its more philosophical justifications. While, of course, calculus is now less ideological and much more solidly grounded than capitalism, we see the force by which objectification works, allowing objects to think for us: you can believe in calculus or not, but while Berkeley was busy debating its justification, Isaac Newton and others were busy solving problems, proving the force of these mathematics.

Berkeley's dereification of calculus prefigures contemporary attempts to dereify algorithmically produced knowledge. He ends *The Analyst* with a prescient question about the difference between knowing and merely computing: "Whether the difference between a mere Computer and a Man of Science be not, that the one computes on Principles clearly conceived, and by Rules evidently demonstrated, whereas the other doth not?"[16] While for Berkeley, the term "computer" meant something drastically different from what it does today, the sentiment is not far afield from Tymoczko's claim that a computationally derived "proof" does not count. Both Berkeley and Tymoczko insist that one must clearly understand the principles and progressions of knowledge, rather than merely apply a set of rules.[17] In this way, Berkeley's concerns about calculus's unsteady foundations pave the way for contemporary unease—such as Tymoczko's—with computationally derived knowledge.

But beyond the insecure ground on which it seemed to operate, what actually made calculus's practitioners "infidels" according to Berkeley? In other writings, Berkeley was forcefully opposed to the

---

16 Ibid., 218.

17 Jacques Derrida's work, especially his later work, in many ways repeats this insistence, claiming that a decision cannot be the application of the rule. See, for example, *Rogues: Two Essays on Reason* (Palo Alto, CA: Stanford University Press, 2005).

beliefs of freethought, a religious-philosophical movement that argued that while God had set the world in motion, its development, laws and nature could be understood through logic and reason.[18] Contrary to the long-held Christian belief in mysteries of faith—such as the existence of the trinity, transubstantiation, the function of prayer, and so on—the rationalist principles of freethought staked its new theology on a logical, and thus humanly knowable, vision of the world.

While Berkeley was an avowed Anglican, his thinking echoed the First Vatican Council, convened by Pope Pius IX in 1868, which reaffirmed the existence of "mysteries." According to the dogmatic constitution that resulted from the council, "divine mysteries, by their very nature exceed the created intellect so much that, even when handed down by revelation and accepted by faith, they nevertheless remain covered by the veil of faith itself, and wrapped in a certain mist, as it were, as long as in this mortal life we are away from the Lord, for we walk by faith, and not by sight."[19] The council continued by declaring that "if any one say that in Divine Revelation there are contained no mysteries properly so called, but that through reason rightly developed all the dogmas of faith can be understood and demonstrated from natural principles: let him be anathema."[20]

For both Berkeley and Pope Pius IX, mysteries cannot be explained through reason and human-divined natural principles. They are opposed to such thought, believing instead in a deistic structure of knowledge indiscernible to humanity's cognitive capacity. Berkeley's criticism of calculus centered on a desire to keep these mysteries of faith mysterious, which he sought to do by demonstrating that calculus is likewise founded on mysteries rather than reason. He asks: "Whether Mysteries may not

---

18 George Berkeley, *Alciphron; or The Minute Philosopher: In Focus* (New Haven: Increase Cooke & Co., 1803).

19 Quoted in Heinrich Joseph Denzinger, *The Sources of Catholic Dogma*, trans. Roy J. Deferrari (Fitzwilliam: Loreto Publications, 1955).

20 First Vatican Council, 3rd Session, "De fide et ratione," i.

with better right be allowed of in Divine Faith, than in Humane Science?"[21]

The aim and target of Berkeley's essay was likely not, as some have suggested, the famed Isaac Newton—who himself was a man of faith—but rather several less-famous mathematicians who rejected the existence of religious mysteries in favor of a wholly rational world, a freethinking intellectual maneuver that declared God's workings profane. In this way, they posed a double threat to the reified theology of the church. On one hand, freethinkers openly attacked the existence of mysteries on account of their irrationalism. On the other, freethinkers tamed the very concept of mystery for the purposes of advancing mathematics and the sciences.

Perhaps freethought would not have been so threatening to Berkeley's worldview if it merely did its work, advancing reason, logic and mathematics while leaving the question of mysteries to the church. Yet for Berkeley, such infidel mathematics created and worked on its own mysteries, and thus established a new theological episteme interior to its own workings. The success of calculus emboldened freethinkers to extrapolate from their mathematical successes to attack the mysteries of the church. Berkeley's argument claimed the necessity of religious mysteries by proving that even freethinkers work by way of mysteries. Thus, in his thought, either the mysteries of calculus undermine its secular foundation of reason, or mathematicians should leave religion to its own theology.[22]

The mysteries that ground both Berkeley's reading of calculus and Christian doctrine are fundamentally opposed to reason precisely because they make the unreasonable claim that *inequality is equality*. The mystery of the trinity relies on the mathematically

---

21 Berkeley, *The Analyst*.

22 Cantor argues that *The Analyst* should be understood as a demonstration that calculus partakes of the same mysteries as religion, which for religion is foundational but for calculus is an unbearable affront to its rationalist claims. Geoffrey Cantor, "Berkeley's *The Analyst* Revisited," *Isis* 75, No. 4 (1984): 677.

impossible formula of 1 = 3. Transubstantiation declares an equality between bread and the body of Christ. And the "ghosts of departed quantities" of calculus's infinitesimals appear likewise as a paradox: the equality, and coterminous possibility, of presence and absence. All of these mysteries trouble the rationalist ground of knowledge while simultaneously providing its building blocks.

Such mystery lies at the heart of all objectification. In the case of calculus, it is these ghosts. In the case of capitalism, for Marx, it is the act of making unequal types of labor and commodities equal that founds the objectification of capitalist exchange. And, in the case of computation, it is a process of comparing, and thus calculating, seemingly incommensurate information about the world. Such mysteries cannot be negated simply because they are seemingly spurious, for as long as those mysteries "work"—economically, materially, theologically—the simple act of showing what they are "really doing" will never be enough to disempower them. In any revolutionary politics, it is a necessary step to reveal how a mystery or inequality functions; but such a politics also requires the effectuation of new mysteries—just as calculus posited its own.

Epistemic Authority

Revolutions in mysteries and metaphysics, whether of labor or knowledge, have a tendency to call into question old mysteries and the high priests who proclaim them.[23] The freethinkers with whom Berkeley clashed favored logic over authority and believed in the possibility of a fundamentally clear grasp of the world—one that would attempt to do away with authority altogether. In this sense, it was consistent with many other revolutionary modes of thought, including those data scientists, who believe that ability to work with

---

23 In regard to probability, see Ian Hacking's excellent text, *The Emergence of Probability: A Philosophical Study of Early Ideas about Probability, Induction and Statistical Inference* (Cambridge, UK: Cambridge University Press, 2006).

data can supplant the authority of domain knowledge.[24] James Jurin, a freethinker and contemporary of Berkeley, attacked Berkeley for appealing to the general reader to decide whether his claims are justified. He asks, "Pray, Sir, who are these thinking readers you appeal to? Are they Geometricians, or persons wholly ignorant of Geometry? If the former, I leave it to them: if the latter, I ask, how well are they qualified to judge of the method of Fluxions?"[25]

In response, Berkeley admonishes Jurin for his autocratic reliance on Newton's expert authority and insistence that the general reader cannot judge his mathematical arguments: "In a matter of science, where authority has nothing to do, you constantly endeavour to overbear me with authorities ... No great name on earth shall ever make me accept things obscure for clear, or sophism for demonstrations. Nor may you ever hope to deter me from speaking what I freely think."[26]

While we should likely give Newton his due and respect his expert insights, Berkeley catches his opponents in a difficult dilemma: If we favor reason over authority, how are we to judge the other's reason where we have not reasoned it out ourselves? But this dilemma, and the crises of authority to which it speaks, tends only to move in one direction. The freethinkers and computers appear not to need the force of authority (even if they accidently call on it here or there), because the objective nature of their claims no longer requires a locus; it grounds itself in its own groundlessness. In appearing to work, it finds itself in the appearance of the world; it objectifies itself as knowledge.

The effort to establish the validity of these mysteries—which make the incommensurate commensurate either by authority or by

---

24 Along these lines, Lewis Mumford argues that the ability to work with glass played a critical role in the rise of a thought committed to clarity: *Technics and Civilization* (Chicago: University of Chicago Press, 2010).

25 James Jurin, "Geometry No Friend to Infidelity," *London* 1734 (2002): 382.

26 George Berkeley, "Defence of Free-Thinking," cited in Cantor, "*The Analyst* Revisited."

the objectified appearance of some force without place—ultimately amounts to the same thing: the provision and assumption of a ground for thought. These knowledges of new mysteries—just like the probabilities that an algorithm produces, the value of commodities or the infinitesimal—always rely on some non-present ground as their foundation, in a process that says, "Let us act and think as if this were true." But despite this mysterious ungroundedness, they still function as if they were real—as ghosts of departed quantities—or, as Alfred Sohn-Rethel calls them, real abstractions.

It is precisely through these mysteries that the world is made "real," from the consecration of moral knowledge through divine intervention to feats of engineering fabricated from integrals in calculus. And once these mysteries are made real, succeeding in their feats, they begin to work on their own accord; in short, they appear objective. They cease to be solely a means by which we count—whether the trinity, mathematical proofs, the price of goods or acceleration—and they begin to account for us and our affairs. In short, these mysteries become objective, and allow us to reason and interpret the world.

In the end, it is folly to simply and exclusively deny the existence of this process, because it is through mysteries and their linkage of equality and inequality—or, to put it differently, abstractions, like class, race, gender, nation and value—that the social world is produced and acted upon, as the detritus of history is drawn into new forms. But certain mysteries—namely those that have supported injustice and exploitation in the form of capitalism, imperialism, racism, sexism and so on—must be undone and their power resisted. To simply ignore them—as we saw in the first chapter with Chris Anderson's glorification of correlationism—is to allow them to continue to function. Thus, we must repudiate Berkeley and his epistemic anxieties toward new mysteries. Instead, the task is to ourselves become revolutionary, or in Berkeley's terms infidel, mathematicians.

Mysteries serve to provide a metaphysical ground and force to these historical constructions. Under capitalism, as philosopher

Moishe Postone has argued, these mysteries become fully relativized, removed from any transcendental grounding in God, the state or even the scientist who knows; their ground loses any determinable locus. And with algorithms, probability and statistics, these non-locatable mysteries enter even the production of knowledge. But they do not break with the past; on the contrary, they reproduce it in ever more insidious forms that refuse being undone even by those who may succeed in showing how things really work.

We must recognize the mysteries of reason, the unreasonable ground of reason, and thus the unequal equalities that permit computation and calculation. But recognition of this is not an argument against reason, or against calculation; rather these mysteries are reason's very raison d'être. To understand computation, we must recognize the process by which unequal elements are made equal—and hence accountable. Today, this process holds us to account by capital, which seeks to make all materiality and knowledge commensurate as value. If we are to move beyond the injustices of these systems, both economic and algorithmic, we must ultimately, like the infidel mathematician, propose new mysteries that can build bridges, trouble old mysteries and undo archaic ideologies.

Chapter 3
Algorithms of Objectification

While the general importance of algorithms may appear clear, a proper definition of the term "algorithm" itself is far from settled and has been an ongoing area of debate for over half a century. In his famous text *The Art of Computer Programming*, computer scientist Donald Knuth offers a definition based on five requirements: (1) finiteness—an algorithm runs and finishes all of its steps in a finite amount of time; (2) definiteness—each step is rigorously and unambiguously defined; (3) input—an input is taken, and through this process produces (4) an output; and (5) effectiveness—each step must be simple enough that it can easily be understood and "be done by someone using pencil and paper."[1] While these five requirements have not fully ended the debate about what an algorithm is, they provide a helpful starting point.

From Knuth's definition, two important elements should be noted. First, to "run and finish all of its steps"—or computation—is a process that takes time for the algorithm to complete. While computation is often associated with instantaneity, each step takes

---

[1] Donald E. Knuth, *The Art of Computer Programming*, Vol. 1, *Fundamental Algorithms* (Reading, MA: Addison-Wesley Longman, 1996), 6.

at least a tiny fraction of a second. For complicated algorithmic models that run over large datasets, like the artificial neural network from Chapter 1, computers often struggle to complete their instructions in a reasonably finite amount of time. Yet with the increasing power of computers—meaning faster operations and thus less time needed to process—algorithmic instructions can now operate on new and more complex scales. Theoretically, one could, as Knuth suggests, sit down and compute any algorithm with pencil and paper—including the proof of the 4CT or the mathematics necessary to send a human being to the moon. But there is not enough time in a human lifespan to independently compute the entirety of something as complex as Google's famous PageRank algorithm (even though each individual step could be done by hand). Thus, to operate at this scale, we need computers.

Second, algorithms engage in an act of inscription: an algorithm transforms an input into an output. For modern computers, this means flipping physical bits in order to take a set of symbols (letters, numbers, etc.) and produce a new set of output symbols—a process that mathematician and early computer scientist Alan Turing famously demonstrated with his theoretical model of computation, the Turing machine. In his machine, there is an infinite reel of tape, divided into cells, from and onto which the machine can read and write. With a relatively small table of possible instructions represented as symbols, and the ability to move backward and forward on the tape, Turing proved that it was possible to encode any computer algorithm using this machine. Since this theoretical discovery in 1936, it has been possible to understand all algorithmic instructions as symbol manipulation, and hence as a species of writing. The ability for algorithms to "write" in this way facilitates a new form of memory; thus, this ability to remember and account for our affairs has become one of computers' most consequential attributes.[2]

---

2 See Justin Joque, *Deconstruction Machines: Writing in the Age of Cyberwar* (Minneapolis: University of Minnesota Press, 2018).

In this way, algorithms, and along with them a longer history of capitalist computation from bookkeeping to price settings, are always opaque—not only in that we do not necessarily know how an algorithm arrived at a given result, but also in a deeper metaphysical and epistemological sense, in so much as human operators of computers must trust what has been written to memory at a speed and scale that is necessarily imperceptible. In this world, the only way that computation can effectively function is if it is allowed to manage itself. It is here that data, algorithms, and thus computation present themselves as objective. Without the ability to know what is happening in algorithmic processing, we must trust the results of the output values. Just like the market forces of capitalism, value is computed by complex processes whose totality is necessarily inaccessible and appears to arrive from elsewhere. These objective values are not synonymous with "truth." Rather, a certain truth is attested to—or vouched for—by some object; in short it is objectified.

Tally Sticks and Objectification

At the risk of pushing the bounds of what "counts" as an algorithm, the twelfth-century medieval European use of split tally sticks offers an exemplary instance of how algorithms are able to objectify knowledge. A split tally stick was a device for keeping track of loans and debts, useful in a world lacking both coins and mass literacy. To record a debt, the numerical amount of goods or monetary equivalent was carved into a stick.[3] If a farmer lent another farmer twenty sheep, twenty cuts would be physically etched into the stick. The now-inscribed stick was then broken in half across the inscribed tallies. Each half contained part of each original mark; and both parties were given a half of the stick as a

---

3 W.T. Baxter, "Early Accounting: The Tally and Checkerboard," *Accounting Historians Journal* 16, No. 2 (1989): 43–89.

receipt for the transaction. The possibility for fraud was structurally precluded: the debtor was unable to erase any of the marks, since they were cuts into the wood, while the lender was unable to add any additional debts to the count, as they would only appear on his half of the stick.

In the parlance of modern computer science, the use of split tallies sticks leveraged the immense "search space" of wood—search space being the number of possible solutions to a problem, such as all the possible four-digit codes to unlock a cell phone. One of the only ways someone could cheat the system (short of stealing their rival's physical stick) was to find a similar stick, carve fewer notches in it, and break it exactly the same way. The sheer number of types of sticks—both in terms of wood, but also size and growth pattern—and the near-infinite number of ways that a stick can break, constitute the search space of the problem: the number of possible sticks and breaks one would need to search through to find a suitable replacement to cheat the system. The tally stick algorithm "worked," and in working, it provided something vital for the economies of that time: a physical object that could account for commercial affairs.

The efficacy of the split tally stick's method of accounting attests to the power of objectification. As with the definition of algorithms themselves, there has been a long and healthy theoretical debate about how this concept plays out. For our purposes, Marx's description of the process of objectification, especially in the first volume of *Capital*, is critical to understanding algorithms, machine learning and statistics.[4] At its heart, objectification is a theory of

---

[4] In a somewhat famous footnote, Louis Althusser objected to what he called "the theory of reification," suggesting that it projected the early theory of alienation onto the latter theory of fetishism, creating a concept that he believed was too psychological. Regardless of whether one accepts Althusser's periodization of Marx's thought, the purpose of objectification as developed in this book is to provide a theory that is decidedly not psychological, even though latter chapters engage with the concept of alienation. The point here is that objectification creates an external, though non-localizable, force of necessity. Althusser, "Marxism and Humanism," in *For Marx*, 230n7.

how objects think, remember and work for us. For Marx, objectification is not, as in contemporary idiom, the process of treating a person like an object; rather, he defines the term as using objects to manage human affairs.

One experiences this process of objectification acutely when airline representatives look at their computer screen after a canceled flight and sympathetically say, "Sorry, but there is nothing I can do; the computer won't let me change anything." One can decry the situation all one wants, but it is clear the system has been built such that even the most persuasive of critiques will not book a different flight. Tellingly, airline representatives are no longer called gate agents; they have surrendered their agency to the computer system.

This process is precisely what lies at the heart of algorithmic power: computation presents us with an "objective"—and hence politically unassailable—description of a given sociopolitical situation. The force of these objective facts inheres not in their correspondence with some external reality, but in their ability to directly produce a new, distinct social reality. For example, what matters most about credit reports is not their ability to accurately describe an individual's creditworthiness, but their ability to produce a world that corresponds to their priorities by limiting access to capital; indeed, it is telling that they are increasingly factored into hiring decisions.[5]

At its most fundamental level, the process of objectification serves two interrelated purposes: first, it allows an object to remember for us; and second, in doing so, the object also *believes* for us. This activity of memory and belief is simultaneously quotidian and nothing short of magical.

With the tally stick, the notches remember for our two peasants exactly how many sheep changed hands. And because of the

---

5 Barbara Kiviat, "The Art of Deciding with Data: Evidence from How Employers Translate Credit Reports into Hiring Decisions," *Socio-Economic Review* (2017).

ingenious system of splitting the tally across two sticks, it becomes possible for both of the peasants to believe what the stick remembers for them. In fact, this technology became so effective that King Henry I instituted its official use for tax collection in 1100, a practice that was kept in some use throughout England until the early nineteenth century.[6] Moreover, due to its economic importance, Henry I's edict endowed these sticks with the force of law. It became not only easy, but functionally necessary, to believe what the sticks remembered.

Just like a computer, these tally sticks function as memory devices. Our peasant can safely put the number of sheep he lent out of his mind, for it is inscribed in the stick. Also like computers, they aid in computation. With multiple sticks one can easily add, subtract and account across multiple transactions. Moreover, they are effective tools for communication. If our peasant were to pass away, his heirs would have little trouble understanding the exact amounts of his various debts. Debt thus becomes objective, carved into the material world. As a result of the tally stick, one no longer needs to believe or remember; it is there in the world for anyone to see.

Contemporary understandings of objectification have strayed slightly, but in a theoretically important way, from this reading of Marx. Georg Lukács is partially responsible for this wandering with his interpretation of objectification—or reification, as the term is often translated with his writing—as the point when "a relation between people has taken on the character of a thing."[7] Rather than focus on the process by which objects manage our affairs—even if the outcome is that relationships are taken as objects—Lukács is concerned with the outcome of this process and what it occludes. On this level, his conclusions are half right. Objectification *is* obfuscatory, but what is obfuscated is not fundamentally that people or relations are treated as objects, but rather

---

6 Baxter, "Early Accounting."
7 Lukács, *History and Class Consciousness*, 83.

that objects are made to think in the place of people. This strand of thought moves even further from Marx when, as is common today, objectification comes to mean that individuals, rather than relationships, are treated as objects.[8] These theoretical developments are not necessarily wrong, and in many ways they are the outcome of capitalist objectification, but still they risk overlooking the processes through which objects account for social relations.[9]

Thus, while many commentators present objectification as a type of "false consciousness," where one forgets the true social relations that underwrite, process and thus muddy even the most clear-sighted analyst's mind, it should be clear from the example above that, for present purposes, objectification can be read as significantly more neutral.[10] Objectification is a form of forgetting, but one that is directly productive. It allows objects to relieve us of the necessity of remembering, an unburdening of our minds that is sometimes beneficial.[11] For any objectification, the question will always be who is burdened, who is unburdened. This

---

8 Martha Nussbaum, *Sex and Social Justice* (Oxford, UK: Oxford University Press, 1999).

9 An important distinction must be drawn between the theory of the revolutionary object presented here and Latour's work on Actor Network Theory which attempts to describe the social force of objects—or actants as he calls everything. His theory focuses on the means by which various objects act, whereas the Marxist reading of objectification foregrounds the means by which social relations are inscribed—even if magically or metaphysically—into objects of accounting (e.g., the commodity). Bruno Latour, *Reassembling the Social: An Introduction to Actor-Network-Theory* (Oxford, UK: Oxford University Press, 2005).

10 Georg Lukács, *History and Class Consciousness: Studies in Marxist Dialectics*, trans. Rodney Livingstone (Cambridge, MA: MIT Press, 1967); Herbert Marcuse, *One-Dimensional Man: Studies in the Ideology of Advanced Industrial Society*, 2nd ed. (London: Routledge, 1991).

11 Axel Honneth claims that reification is a form of forgetting, but for him it is the forgetting of an "antecedent recognition" that first allows one to recognize the subjective qualities of an object. With reification, human beings fail to acknowledge the multiplicity of "subjective conceptions and meanings" that individuals accord to objects in their initial recognition. *Reification: A New Look at an Old Idea* (Oxford: Oxford University Press, 2005), 63.

unburdening is the essence of all computing—and all recording and communication technology. From the first etched clay tablet to the latest $200 million supercomputer, all of these technologies remember our affairs for us. And, in remembering for us—our transactions, our social relations—they hold us accountable to the traces of these interactions. While Marx lays out a theory of objectification in terms of value, we can see the same process and force at work in computation.

Commodities and Objectification

In the opening pages of *Capital*, Marx launches his entire theory of capitalism with the simple commodity and its powers of objectification. Everything in his analysis expands outward, from the relations between commodities themselves to relations between laborers and capitalists. Marx proposes a radical understanding of objects-as-commodities: what is objective about these objects is not that they are useful, but rather that the social relations between producers, labor and purchasers are congealed into the objects themselves. Commodities "hide" the very relations and histories that determine their value as they come to market.

Marx, following the economists of his day, argues in these early pages that what determines the economic value of an object is not how useful it is—some of the most useful things in the world are the cheapest—but rather the amount of labor that goes into its production. Accordingly, economic value is not inherent in the object. It is, rather, a reflection of the entire social process of production, work, banking, and so on. For Marx, "a use value, or useful article, therefore has value only because abstract human labour is objectified or materialized in it."[12] In this way, capitalism, in essence, operates already as a giant distributed and opaque algorithm, computing and remembering the amount of labor that

---

12 Karl Marx, *Capital*, Vol. 1 (London: Penguin Books, 1990), 129.

should go into the production of any product, then presenting the output of this calculation anytime we go to buy something. While it may seem irrational to believe that physical objects have economic value as an intrinsic property, capitalism compels one to act as if they do.

Thus, for Marx, these objects serve an immensely important role in so much as they record social relations and then present them back to us as the objective state of things. In his words, "The object which labor produces—labor's product—confronts it as something alien, as a power independent of the producer. The product of labor is labor which has been embodied in an object, which has become material: it is the objectification of labor."[13] The syntax here is elucidating: it is the object that is objectified. As soon as labor is poured into a commodity, the commodity appears to hold onto that expended labor in the form of economic value, carrying it wherever it goes regardless of the wishes or beliefs of its producers or owners.

Moreover, this commodity-as-object is fundamentally computational and networked; it relates all laborers and consumers to each other. Marx further says of the commodity:

> A commodity may be the outcome of the most complicated labour, but through its *value* it is posited as equal to the product of simple labour, hence it represents only a specific quantity of simple labour. The various proportions in which different kinds of labour are reduced to simple labour as their unit of measurement are established behind the backs of the producers.[14]

Here, Marx presents an equalization of labor that ultimately lays the theoretical foundations for his analysis of the mechanisms of

---

13 Karl Marx, "Economic and Philosophical Manuscripts of 1844," in *The Marx-Engels Reader*, Robert C. Tucker, ed., 2nd ed. (New York: W.W. Norton & Company, 1978), 71.

14 Marx, *Capital*, 135.

capitalism itself. For Marx, all economic value is based on the accumulation of simple labor: the amount of time it takes an average unskilled laborer using available technology to produce a commodity. While competing theories of value exist, any rigorous concept of value will necessarily "reduce" a variety of specific practices of labor to a single economic value—a value that can be compared to other commodities, which each have their own value. In this way, all value is relative only to some other value.

For example, an amateur could spend ten painstaking years learning, failing and testing to produce a light bulb of equivalent quality to a generic supermarket brand. Despite the difference in production, and short of convincing consumers that the amateur's repeated folly adds some artisanal patina, that bulb will be valued equal to the supermarket bulb created in mere minutes. In this determination of value, the market thinks for consumers, reducing all value to a network effect that somehow—as if by magic—computes the amount of labor it *should* take to make a commodity. Moreover, this computed value is itself a function of how much it costs for laborers themselves to live and reproduce. Like an algorithm, the market computes a solution to the problem for use in current and future decisions.

The commodities that labor produced come to account for human affairs "behind the backs" of workers. The bulb's value, in relation to all other value, determines not just how much is paid by consumers, but the entire network of value that points only to itself. This commodity-as-object emerges as the agent that structures capitalist society: the object itself is a revolutionary force. Like a globally distributed social computer, the wide array of commodities across the capitalist marketplace tracks the value of all of the labor that goes into their production, constantly updating as conditions change. In a sense, the commodity has the whole universe folded inside it. Commodity exchange, by allowing the real-time calculation of these ratios, in essence becomes a form of "machine learning" in a much broader sense than the term usually connotes.

To see this in action, we need look no further than the capitalist invention par excellence: the stock market. In this market, traders attempt to divine what others will believe the future value of a company to be. What is fully a human process of production and exchange appears, to the trader, as an unknowable and fickle system whose whims are even less predictable than the weather. Listen to the headlines on any given day: "Stocks retreat from records after winning streak"; "Stocks limp to the end of another winning week." Both journalists and readers should know better—stocks are not freely acting agents and thus do not "retreat" or "limp." But they, and we, still act *as if* they are, and do. Stocks appear as objective; they confront us and, regardless of whether we believe in them or not, require that we come to terms with their reality.

Prefiguring the logic of computation, objects function on a plane of formal equality, one where even the most unequal is made computable. Like a stock market list of a thousand different firms' acronyms, the functional, qualitative differences between each corporation are collapsed. By reducing all exchange to the capitalist conception of value, all types of objects are made equal, or at least equatable. Moreover, once this process is underway, it carries on without our consciousness involvement: "They do this without being aware of it."[15] In essence, the objectification of each commodity collapses value, and with it, labor, onto the same scale. It makes the resulting object computable such that one no longer even needs to be aware of how that object came to be. Commodities do this magical work by creating the appearance that value is somehow internal to the object itself. And so, value too appears as a "ghost of a departed quantity" of the labor that produced it.

Objectification has prepared the world for its computation, and the contemporary world, managed by information capitalism, statistics and algorithms, has taken up this mantel of objectification. Though they clearly have not replaced the objective power of

---

15 Ibid., 166–167.

wage labor and commodities, these computational forces manage, maintain and reshape those capitalist processes.[16] Algorithms track, process and account for significant portions of financial, social and scientific interactions, while statistics provides the metaphysical ground of these systems, explaining and justifying how individual interactions can be combined and computed into social truths. In short, statistics provides the objectifying force of algorithmic knowledge and the equatability necessary for its computation, just as labor provides the force for capitalist value production; and labor's reduction to value provides the grounds of its computation in the great social network called the market.

Objective Metaphysics

Objectification is ultimately a metaphysical process. It sutures a metaphysical structure to the material world, where individual commodities—as well as algorithmically produced outputs—"transcend" their particular conditions in order to connect to a whole networked world of others. Marx describes the object as "abounding in metaphysical subtleties and theological niceties."[17] He further makes the theological nature of this process explicit, stating:

---

16 While Marx was writing about a heavily industrial economy, many have attempted to update these insights to describe contemporary capitalism. For example, in a 1981 chapter of his book *Dependency Road*, Dallas Smythe reinterprets Marxist critiques of political economy to account for the role of advertising and audiences in contemporary capitalism. Smythe describes audiences as commodities that mass media industries produce as consumers whose attention they sell to advertisers. *Dependency Road: Communications, Capitalism, Consciousness, and Canada* (Norwood, NJ: Ablex Publishing, 1981). Christian Fuchs has updated Smythe's theory of the "audience commodity" to think about how social media companies like Facebook and Twitter extract value from users. For more, see *Digital Labour and Karl Marx* (London: Routledge, 2014), esp. 72–134.

17 Marx, *Capital*, 193–194.

There it is a definite social relation between men, that assumes, in their eyes, the fantastic form of a relation between things. In order, therefore, to find an analogy, we must have recourse to the mist-enveloped regions of the religious world. In that world the productions of the human brain appear as independent beings endowed with life, and entering into relation both with one another and the human race. So it is in the world of commodities with the products of men's hands.[18]

Objectification is, in a way, even religious (which we see again when Marx uses this process as the grounds of commodity fetishism).[19] The object performs the mysterious and metaphysical work of equating its own individual being with a global world of exchange, maintaining the relationship between use value (what is useful about a commodity) and exchange value (what it is worth on the market).[20]

Just as with religious mysteries, once the machine is set or once the theology is instantiated, any new analysis can only produce results that confirm the system. A dip in a stock price suggests a company is in peril. A new scientific discovery can be integrated into God's plan for the world. Belief in objects' value begets belief in the system, a vicious cycle that complicates any politics founded on an apprehension of the world as it really is. As soon as the object and its value are naturalized (e.g., the tally stick is agreed to

---

18 Ibid., 165.

19 To Moishe Postone, all three volumes of *Capital* represent attempts by Marx to draw attention to the way fetishism conceals the abstract dimension of labor through forms that appear as transhistorical and metaphysical. *Time, Labor, and Social Domination* (Cambridge, UK: Cambridge University Press, 1993), 271. For a nuanced reading of the notion of fetish, especially as it was historically developed from Afro-Atlantic gods, and the social rationality that underwrites its various valences, see J. Lorand Matory, *The Fetish Revisited: Marx, Freud, and the Gods Black People Make* (Durham, NC: Duke University Press, 2018).

20 Marx says: "The internal opposition between use-value and value, hidden within the commodity, is therefore represented on the surface by an external opposition, i.e., by a relation between two commodities." *Capital*, 153.

represent a certain debt), the game is set. Objectification produces the default; it delimits the world by defining the relevant statistical reference classes or the stakes by which engineers and corporations will determine whether an algorithm has succeeded.

It is for this reason that attempts at dereification, to show what is really going on, are so challenging, for they tend only to show what everyone already knows is ridiculous. Marx writes: "Value, therefore, does not have its description branded on its forehead; it rather transforms every product of labour into a social hieroglyphic. Later on, men try to decipher the hieroglyphic, to get behind the secret of their own social product."[21] Many progressive critics have followed this road of hieroglyphic interpretation, tending not to seek the nature of the system producing value but rather only the value of this or that product, condemning all subsequent analyses to a *post festum* moment in which it is too late to question the underlying structure of the social world.

These attempts at dereification take up the task of unearthing what has been entombed in the seemingly natural ground beneath one's feet, focusing solely on the meaning of what has been buried, rather than on the process. Once the mold of objectification is cast, and objects begin to handle human affairs (which then become the objects' affairs), it becomes nearly impossible to arrive at a conclusion outside the hard-coded presumptions set in that mold. Even when one can see through the haze of these objectified relations, they still have to act as though they are true. These objects come to underwrite the form of their own relations, as in the case of money: it is obvious that a piece of paper is not worth anything, but it is still imperative to act as though it is. Pointing out its worthlessness will change very little.

---

21 Ibid., 167.

Objectification: Not a Blueprint

Objectification, as it is understood in this sense, is not an ideology in the strict sense, nor is it the totality of politics. It is decidedly not a misidentification of the social world whose correction would guarantee the end of all injustice. Rather, it functions as the means by which ideology, politics and social relations in general are given a non-locatable force. In his book *Black Marxism*, political scientist Cedric Robinson argues for "the nonobjective character of capitalist development," giving the first chapter on racial capitalism this subtitle. By this he means that capitalism does not follow some definitive laws of historical or dialectical development, especially not those that would see capitalism increasingly rationalize production and strip away old prejudices of race, gender and nation. Likewise, abolitionist scholar Jackie Wang, in *Carceral Capitalism*, shows how the predatory violence of the modern neoliberal state, exemplified by the United States, where mass incarceration, especially of Black men, constitutes one of the "gratuitous forms of racialized state violence that are 'irrational' from a market perspective."[22] It is these various forms of violence, from enslavement to mass incarceration, then, that mark the nonobjective nature of development. Robinson says, of the "development of revolutionary consciousness among Black and other Third World peoples" that it "broke with the evolutionist chain in, the closed dialectic of, historical materialism."[23]

Is it still possible then, in this light, to speak of objectification, especially if we accept the nonobjective nature (or at least element)

---

[22] Jackie Wang, *Carceral Capitalism* (Cambridge: MIT Press, 2018), 22.

[23] Cedric Robinson, *Black Marxism: The Making of the Black Radical Tradition* (Chapel Hill: University of North Carolina Press, 2000), 276. Fred Moten asks of this quote: "Is it a complete detachment from that temporal/historical trajectory or is it a displacement, a retemporization disruptive of the very idea of absolute break and, by extension, an augmentative curvature of old harmonic notions of convergence or hybridity, a dissonant bending of the dialectic and its notes?" *Black and Blur* (Durham, NC: Duke University Press, 2017), 9.

of capitalist development, or in Wang's terms, a sadistic element to white supremacy (and also to gendered violence) that does not fit, or outpaces, economic analysis?[24] For this precise reason, it is important not to understand objectification as a form of false consciousness—as a misguided perception that can be rectified, and that (as certain arguments for historical evolution argue in the case of racism and sexism) capitalism will or could rationalize away. In this way, objectification would define not a rationalization of domination and violence, but rather its abstraction, its loss of a determinable locus, to invoke Postone's work from earlier. Thus, objectification is not a rationalization, as the term is normally understood, but the process by which concrete domination (e.g., racism, sexism, imperialism and class exploitation) is translated into an abstract form whose origination and social elements appear to recede behind its objective mask.[25]

Growing concerns about YouTube's video recommendation algorithms provide an example of this dynamic. A number of commentators have documented how the recommendations tend to push viewers toward reactionary conspiracy theory content, especially toward far-right, white-supremacist and misogynist videos, making YouTube into what effectively amounts to a radicalization machine.[26] The algorithm is not explicitly designed to push increasingly reactionary content, but rather to encourage users to keep watching in order to sell more ads. It "just happens" that the algorithm has discovered the most efficacious way to do this is to radicalize viewers; in short, it follows the "objective" laws of the market—attempting to sell as many ads as possible. As scholar of Black studies and digital culture Ramon Amaro

---

24 Wang, *Carceral Capitalism*, 89–92.

25 It is possible to see this process in Marx's own analysis of imperialism in the third volume of *Capital* and the significant secondary literature that builds upon it.

26 Zeynep Tufekci, "YouTube, the Great Radicalizer," *New York Times*, March 10, 2018; Kelly Weill, "How YouTube Built a Radicalization Machine for the Far Right," *Daily Beast*, December 17, 2018.

describes such systems, "What we experience today as algorithmic prejudice is the materialization of an overriding logic of correlation and hierarchy hidden under the illusion of objectivity."[27] Thus, objectification processes the subjective domination inherent in the system, constantly seeking new ways to extract value from it and thus intensify it. YouTube, in following these objective laws, reproduces and intensifies racist and sexist content, further increasing its potential to produce profit from this content in the future. The logic of ad sales thus provides the form and force by which white supremacy and other far-right ideologies, including their often-sadistic "nonrational" elements, are reproduced and incentivized.

This decidedly does not mean that racism and sexism are accidental to capitalism in general, or, in this case, to the sale of advertisements. Indeed, these forms of domination allow the very production of profit in this instance, and, in the broader case of capitalism, support various forms of theft, exploitative and forced labor conditions and nationalist fear that capitalism requires for its reproduction. The point is that while racism and sexism are made to function in certain ways, made abstract though the drive for profit, they are not reducible to capitalism. Moreover, this does not imply that YouTube's algorithm should not be resisted or objected to. In fact, it is through criticisms of such instances that it may be possible to envision some other objectification that could provide further ground for resistance.

While the roots and causes of discrimination may lie before and outside of capitalist development, the "objectivity" of the market gives this violence a specific form that at the same time makes it increasingly difficult to locate its ground. Moreover, by further segregating populations, both physically and digitally, and preventing certain people from having access to space and capital, capitalism reinforces and economically incentivizes concrete domination—which, in being objectified, interlocks these concrete and abstract forms of racial and gendered violence. In short,

---

27 Ramon Amaro, "As If," *e-flux*, February 14, 2019.

objectification becomes the means through which concrete domination and violence are given abstract form and then translated back again into the concrete; only now, this occurs with a force whose origin appears absent. We shall see that this is precisely how Bayesian statistics functions: whether or not the outcome entails violence, it provides a method for converting subjective beliefs about the world into probabilities that have objective force. In short, its proponents' mantra will be: "Individuals can believe whatever they want, but we will show you how to produce value from those beliefs."

From this, two key points should be noted: First, objectification is not totalizing, even as it shapes and interacts with a multitude of forms of violence. Sadistic forms of concrete violence still operate and drive society in unobjectified form and have clearly locatable perpetrators (e.g., those that appeared during the Trump administration, and other resurgent nationalisms worldwide). Second, while the present text focuses on the metaphysics of objectification, abstract domination cannot be separated from concrete domination. The abstractions that are at stake are "real abstractions," in that they depend on material conditions and directly bear on life and the ability of individuals to live. Moreover, even though concrete domination cannot be reduced to market incentives, such domination serves to provide a variety of means for expropriation that directly produce value, from private companies that profit off of imprisonment to the exploitation of underpaid labor in the global South. Thus, any resistance to capitalism cannot succeed solely on the level of the abstract but must simultaneously change concrete conditions.

Bitcoin

The function of objectification can be seen clearly in the twenty-first-century update of the tally stick: the blockchain. Developed as part of the Bitcoin digital currency algorithm, the blockchain has become the backbone of numerous attempts to "objectively" account for online transactions. Run through tens of thousands of different computers across the globe, this distributed blockchain acts as an algorithm that automatically accounts for the entire Bitcoin system.

Bitcoin's algorithm confirms transactions by leveraging the lengthy periods of time it takes to compute cryptographic functions, unless one has the secret key. By rewarding "miners" with slivers of its currency to use these significant amounts of computing time to confirm transactions, the blockchain, in theory, is able to automatically confirm transactions and take care of itself, removing the need for any central monetary authority.[28] It is through this "proof of work," in essence the use of extensive processing time and hence electricity to search through possible solutions, that the ghost of this departed work "objectively" appears to mark some amount of value.

The advancements heralded by Bitcoin's technology have led many to see a revolutionary potential as the computerized maintenance of the blockchain removes the need for human oversight and even agency; for instance, a recent group of activists has advocated what they call "cypherpolitics"—an approach that supposedly leaves the entire notion of belief behind. One of its exponents, Stacco Troncoso of the P2P Foundation, writes: "There is no human way of knowing if someone has expressed the truth. This can only be verified through technology. The only way for someone to subscribe to a cypherpolitics is to leave all traces of belief systems behind and only maintain the absolutely essential approximation of the 'truth' . . . By replacing physical trust quotas with immutable code, the blockchain resolves this issue."[29]

For cypher activists like Troncoso, it is the immutability of

---

[28] See Finn Brunton, *Digital Cash: A Cultural History* (Princeton, NJ: Princeton University Press, 2019).

[29] Stacco Troncoso, "Cypherpolitical Enterprises: Programmatic Assessments," *P2P Foundation*, March 15, 2017.

code—like the immutability of a broken tally stick—that allows technologies like blockchain to stand in for belief. Yet, in order for the blockchain to truly function, like any commodity or representative of value, one still must believe in the objectified form. It is thus unlikely that the development of cryptocurrencies will result in revolutionary change, since they are ultimately founded on the same objectified politics of contemporary capitalism. From scratches on a piece of wood to a secret string of numbers and letters, both tally sticks and blockchains are systems of economic management that are also systems of objective belief—representations that are treated as equatable with value. When one accepts this belief, or labors under it even without believing, one remains locked into the world of commodity exchange, watched over by the ghost of a departed computation. Even the realization that the blockchain guarantees nothing changes very little, as long as others believe in its value.

Likewise, statistics—the operative logic by which machine learning algorithmically produces knowledge about the world—is sanctified with similar epistemic weightiness, objectified by its seemingly mysterious ability (in the sense outlined in the preceding chapter) to equate particular data with universal laws. It is through statistics that social facts, presented as data, are objectively turned into metaphysical truths that stand outside of the physical world and determine our social existence—whether we believe them or not.

Under capitalism, the objectification of the market is disastrous and exploitative, but that does not necessarily mean that the form of this objective forgetting is necessarily negative. It is worth revisiting Marx here—in particular, his *Grundrisse*, where he argues that the objectification of labor in machinery and its rational organization or management allows for a reduction of necessary labor time.[30] Capitalism renders labor increasingly superfluous, a process that could liberate individuals from the labor process, if

---

30 Karl Marx, *Grundrisse: Foundations of the Critique of Political Economy* (London: Penguin, 2005, repr.), 339.

not for the dynamic that animates capitalist society: the abstract drive to accumulate greater and greater quantities of surplus value, which requires the expenditure of labor. As such, the mystery of objectification, according to Marx, is both liberatory, insofar as it makes labor less necessary and allows us to think otherwise, and also exploitative, for it allows the capitalist extraction of greater quantities of surplus labor from the labor process.

These mysteries of objectification, while clearly implicated in the violence and devastation of capitalism (and other systems), still provide for other possibilities. As critic and theorist Fred Moten, drawing on the work of Randy Martin, argues:

> The production of surplus—along with that which it produces and is produced by, "race, class, gender and sexuality as the very materiality of social identity"—has reemerged here with a vengeance. Surplus is the very magic of objects, their fetish character, their mysterious secret. That magic can be terrible . . . But there is also a liberatory force of the surplus, the magic of objects, which we see here in the midst of its very transformation.[31]

It is here that these mysteries distribute the separation between object and subject, between objectification and subjectification, determining, along lines of race, class, gender and sexuality, what counts as surplus value, and what can be expropriated in different ways and with varying levels of violence. For Moten, other possibilities—potentially liberatory ones—lie in this process.

What arrives as dangerous and potentially oppressive in the process of objectification is not its formal structure, but the specific—and unseen—social relations, debts, accounts, transactions and force that hold us to these objectified facts, especially as they become abstract, proprietary and operate at massive scales. The power of this forgetting, and the ability to allow objects to think for us, is not necessarily negative—indeed it may even be full

---

31 Moten, *Black and Blur*, 38.

of possibilities. But for such a program of liberation to be effective, it must not limit itself to the surface effects and means of the metaphysics of exchange, to interpreting what has already been decided; rather, it must change the underlying structures of exchange and objectification, along with the social relations that are left to algorithmic objects. This is the task of a revolutionary mathematics: to create new mysteries, rather than simply attempt to repair those that capitalism has left us.

PART II
The Promise of Frequentist Knowledge

Cease from grinding, ye women of the mill; sleep late even if the crowing cock announces the dawn. For Demeter has ordered the Nymphs to perform the work of your hands, and they, leaping down on top of the wheel, turn its axle, which with its revolving spokes, turns the heavy concave Nysirian millstones. We taste again the joys of primitive life, learning to feast on products of Demeter, without labor.

—Antipater of Thessalonica, on the waterwheel

Chapter 4
Do Dead Fish Believe in God?

Through statistics and probability, algorithmic systems think for us. But in doing so, machine learning algorithms only ever produce probabilistic statements about the world, whether in regard to election results, recidivism rates or the best ads to show a consumer. And, while the mathematics of probability are relatively straightforward and usually calculated through rote computation, the ultimate meaning of probability is notoriously slippery. To understand its force today, one must trace its metaphysics and history through the twentieth century.

In general, people struggle to understand probability. One area of daily life in which this problem is evident is the interpretation of weather forecasts.[1] The greatest confusion often surrounds the issue of what class of events a certain measure of probability refers to. For example, "60 percent chance of rain" is sometimes interpreted to mean that 60 percent of the covered area will receive

---

1 Susan Joslyn, Limor Nadav-Greenberg and Rebecca M. Nichols, "Probability of Precipitation: Assessment and Enhancement of End-User Understanding," *Bulletin of the American Meteorological Society* 90, No. 2 (2009): 185–193.

rain, or conversely that it will rain for 60 percent of the time in question. The correct definition of a probability of precipitation (PoP) is the probability that at any given location within the forecasted area there will be precipitation.

In terms of mathematics, then, the definition of PoP is the probability of any measurable precipitation in the forecasted area multiplied by the percentage of the total forecasted area that will receive rain.[2] If there is a 90 percent chance that some part of the area will receive rain, but only in 50 percent of the area, the PoP is 45 percent. This is because a randomly chosen place has a one out of two chance of being in the precipitation area, and that area has a 90 percent chance of rain. We multiply them together to get a probability of 45 percent.

Yet even with the math sorted, what probability means in practice is not at all straightforward, even among the scientific community.[3] First, as long as the probability is neither 0 percent nor 100 percent, probabilistic predictions are, at their most basic level, unfalsifiable. As climate scientists Ramón de Elía and René Laprise state, "Any person can provide a forecast for the probability of showers for tomorrow by just issuing a number between 0% and 100% without considering the characteristics of the atmosphere. This forecast, which is just a numerical representation of an unfounded opinion, is impossible to prove wrong."[4]

Furthermore, if one is predicting the weather for tomorrow, there remains a stubbornly empirical constant in reality: either it will rain or it will not. There is, in a metaphysical sense, no such thing—outside the complexities of quantum mechanics—as a "probabilistic event." This is what statistician Jerzy Neyman concludes in his famed explanation of confidence intervals: "Can we say that in this particular case the probability of the true value

---

[2] "FAQ—What is the Meaning of PoP?," US National Weather Service.

[3] Ramón de Elía and René Laprise, "Diversity in Interpretations of Probability: Implications for Weather Forecasting," *Monthly Weather Review* 133, No. 5 (2005): 1129–1143.

[4] Ibid.

is equal to α? The answer is obviously in the negative."[5] This pessimism is simply because either the true value is equal to *a* or it is not.

Thus, despite the importance of probability to statistics—and many more scientific fields—no consensus exists on what probability actually means. The American statistician Leonard Savage sums up the nature of probability in regard to statistics well: "It is unanimously agreed that statistics depends somehow on probability. But, as to what probability is and how it is connected with statistics, there has seldom been such complete disagreement and breakdown of communication since the Tower of Babel."[6]

While mathematicians, scientists and statisticians have become proficient at calculating the mathematical equations of probability, the underlying question of what probability itself *means* remains a complex metaphysical question. Mathematician Henri Poincaré argued there are two levels on which these problems function: the metaphysical and the mathematical. "Every probability problem involves two levels of study," Poincaré writes. "The first—metaphysical, so to speak—justifies this or that convention; the second applies the rule of calculus to these conventions."[7] In many ways, the two have evolved and changed together, but the metaphysical interpretation has often defined what is mathematically possible and how to make sense of the mathematics.

In its earliest uses, the term "probable" meant that opinions were attested to on good authority. It was not until the seventeenth century that the term came to mean that a certain event or

---

5 Jerzy Neyman, "Outline of a Theory of Statistical Estimation Based on the Classical Theory of Probability," *Philosophical Transactions of the Royal Society of London*, Series A, *Mathematical and Physical Sciences* 236, No. 767 (1937): 333–380.

6 Leonard J. Savage, *The Foundations of Statistics*, 2nd ed. (New York: Dover, 1972), 2.

7 Quoted in Edmund F. Byrne, *Probability and Opinion: A Study in the Medieval Presuppositions of Post-Medieval Theories of Probability* (The Hague: Martinus Nijhoff, 1968).

outcome was actually more likely than others.[8] Along with this shift in meaning, mathematicians made significant advances in calculations of probability, but they initially focused mainly on events whose outcomes were equally likely, such as a coin flip, a die roll, or drawing a card at random from a deck. In these cases, probability appears as a natural outgrowth of discrete, measurable states. For example, a coin is supposed to land only on heads or tails, and a die roll is always a whole number, normally between one and six. One can calculate these probabilities through elementary division.

This understanding of probability was powerful, and it even enabled calculation of some events that were not equally likely. In these cases, the basic unit of calculation—drawing a single card—was held to be equally likely, and could be used to calculate the probability of an unequal event (the probability of drawing a heart or face card). This method becomes possible precisely because the world of cards is closed, represented by a finite number of fifty-two states from which suits and face cards are distributed in specific and known proportion. While notable advances were made on the calculation of probability between the seventeenth and nineteenth centuries, this classical understanding remained relatively simplistic, and it worked only for these discrete applications. The approach had important conceptual limitations: it failed to account for systems that were not closed, were not physically symmetrical and were not based on initially equally likely outcomes, such as the probability of it being a certain temperature tomorrow.

In light of these limitations, over the course of the twentieth century, and building on work in the late nineteenth century, what became known as the "frequentist" description of probability came to the fore. For frequentists, probability represented something more empirical than ideal: probability was not the ratio of some equally likely event (such as the one-in-two possibility of getting

---

8 Ian Hacking, *The Taming of Chance* (Cambridge, UK: Cambridge University Press, 1990).

heads), but the long-run frequency of a physical system. Under frequentism, if, after hundreds of flips, a coin lands heads 60 percent of the time, the probability of heads would be 60 percent. This new, longitudinal definition opened up the theoretical space for significant advances in statistics and probability theory. Now, frequentists could theorize and quite accurately describe the probability of a system—such as a coin that is not "fair" (e.g., asymmetrical) or the temperature—allowing mathematicians to move beyond the ideal, confining necessity of the "equally likely." The implications of these definitions (probability as equally likely events, versus as frequency over a long run) demonstrate how our understanding of what probability means ultimately determines both the limits of what these calculations are able to do and what meanings can be assigned to the results.

In addition to the metaphysics and mathematics of probability, a third level should be added to Poincaré's architecture of probability problems: the economy. Here it is worth echoing scholars who have claimed that "there is no such thing as raw data"—that the data to which researchers have access directly affects the calculations that can, and should, be made.[9] What data is available and seen as viable, how that data is incorporated into an experiment and, thus, how that experiment produces knowledge is sullied by the economic needs of state, society and scientist as incentives to produce knowledge encourage the overstatement of results, and in the extreme to deliberate fraud. Peer-reviewed academic science production is a particularly illustrative example of the economic dynamics at play.

---

9 Geoffrey C. Bowker, *Memory Practices in the Sciences* (Cambridge, MA: MIT Press, 2008); Lisa Gitelman, ed., *"Raw Data" Is an Oxymoron* (Cambridge, MA: MIT Press, 2013).

A Bug in fMRI

In May 2016, a paper was published in the *Proceedings of the National Academy of Sciences* calling into question a large number of functional MRI studies over the past twenty-five years. fMRI, a tool that allows researchers to see which parts of the human brain respond to various stimuli, has played a major role in modern psychology and brain science. These machines measure magnetic variations in the brain to detect localized variations in the levels of oxygen in blood.

With fMRI analysis, higher oxygen levels correspond to increased neuron activity. Yet sources of noise, like random neuron activity, test subjects moving around, the sensitivity of the testing equipment and so on, can all result in increased oxygen detection that may not necessarily indicate any direct change in neuron activity. Though some scientists may uncritically take the output of such studies as clear expressions of brain activity,[10] each fMRI experiment is intimately dependent on calculations of probability—especially the probability that what is observed is due to some meaningful mechanism rather than chance alone.

In each fMRI experiment, one looks for clusters of concentrated neuron activity. But, given the practical impossibility of avoiding ever-present background noise, it is possible that a set of voxels ("volumetric" three-dimensional pixels) can randomly appear in any fMRI output as a cluster of activity. The more comparisons that are made, the more likely it is that random chance will appear as significant (for the same reason it is very unlikely one single person will win the lottery, but highly likely that someone will win the lottery). fMRI tests contain a huge number of voxels, and the risk of a false correlation increases

---

10 Edward Vul, Christine Harris, Piotr Winkielmen et al., "Puzzlingly High Correlations in fMRI Studies of Emotion, Personality, and Social Cognition," *Perspectives on Psychological Science* 4, No. 3 (2009): 274–290.

proportionally with the number of calculations in a given analysis.[11]

If one runs a million statistical tests on any measurable phenomenon, there arises an almost-guaranteed possibility that something will appear meaningful that is actually not: by random chance, a correlation will be discovered. While traditional statistical techniques for dealing with fMRI data try to mathematically correct for this possibility of error, the paper calling into question years of fMRI research was based on a discovery that most major fMRI analytical software did not sufficiently correct for these errors. In fact, a bug in one fMRI program that overestimated statistical significance had gone unnoticed for fifteen years. These software and methodological errors resulted in a high risk of what is known as "cluster failure," where clusters of active voxels falsely appear as neuronal activity.[12]

The study detailing the high risk of "cluster failures" involved taking data from healthy patients with no associated mental task (e.g., subjects are asked to imagine playing a sport or remember some earlier event). The patients were then divided into groups in order to discover possible differences in brain activities. These researchers used the standard significance threshold of 5 percent (a slippery concept that will be explored more fully in the next chapter), wherein a researcher expects that 5 percent of comparisons between groups would appear to have a significant difference between groups; however, they found an unexpectedly high number of false positives, with up to 70 percent of comparisons in the study having erroneously produced statistically significant results.

After the 2016 paper was circulated, many popular publications published its findings with startling headlines that declared

---

11 Stephen M. Smith, "Overview of fMRI Analysis," *British Journal of Radiology* 77 (2004): S170.

12 Anders Ecklund, Thomas E. Nichols and Hans Knutsson, "Cluster Failure: Why fMRI Inferences for Spatial Extent Have Inflated False-Positive Rates, *Proceedings of the National Academy of Sciences of the United States of America* 113, No. 28 (2016): 7900–7905.

decades of scientific research null and void. The *Guardian* asked, "Has a Software Bug Really Called Decades of Brain Imaging Research into Question?" The *International Business Times* announced, "15 Years of Brain Research Has Been Invalidated by a Software Bug, Say Swedish Scientists." And *ZDNet* proffered the pithy headline: "When Big Data Is Bad Data."[13]

Later, after the shock of methodological failure seemed to wear off, journalists took a far more somber, and scientifically reasonable, approach. These latter articles pointed out that this particular problem only accounts for a smaller subset of fMRI research.[14] One of the "cluster failure" paper's original authors even submitted edits to the initial journal in order to downplay some of the more overstated readings of the paper. The journal declined the changes, perhaps because the more sensational claims were better for circulation and attention.

The debate around this "cluster failure" problem raises a number of important questions around the use of probability in the production of modern scientific knowledge. A fundamental challenge of statistical inference—and with it, much of modern science and knowledge production based on probabilistic assessment—is that one can never statistically know anything "for sure." Rather than making conclusive statements, researchers try to separate between what one can claim is due to chance and what one can claim is due to an actual causal mechanism. But, unlike nonstatistical forms of

---

13 Cyril Pernet and Tom Nichols, "Has a Software Bug Really Called Decades of Brain Imaging Research into Question?," *Guardian*, September 13, 2016; Mary-Ann Russon, "15 Years of Brain Research Has Been Invalidated by a Software Bug, Say Swedish Scientists," *International Business Times*, July 13, 2016; Robin Harris, "When Big Data Is Bad Data," *ZDnet*, July 15, 2016.

14 Robert W. Cox, Gang Chen, Daniel R. Glen et al., "fMRI Clustering and False-Positive Rates," *Proceedings of the National Academy of Sciences of the United States of America* 114, No. 17 (2017): E3370–E3371; Daniel Kessler, Mike Angstadt and Chandra S. Sripada, "Reevaluating 'Cluster Failure' in fMRI Using Nonparametric Control of the False Discovery Rate," *Proceedings of the National Academy of Sciences of the United States of America* 114, No. 17 (2017): E3372–E3373.

knowledge, statistical analyses always face the possibility that what was observed, at officially significant levels, was merely the result of random chance.

Even before the 2016 "cluster failure" paper, prior studies had already begun to suggest structural problems within fMRI statistical methodologies. One of the most poetic examples comes from 2009, when then–graduate student Craig Bennett used a dead salmon as a control subject, placed it in an fMRI machine and proceeded to show the salmon pictures of humans in emotionally charged situations. Bennett then asked the dead salmon to imagine what emotion the human in the image was experiencing. The resultant data, uncorrected for multiple testing, clearly shows a small part of the dead salmon's brain lighting up in response to the images. The most likely explanation for this result is not that the dead salmon was thinking, but rather that random noise picked up by the fMRI appeared statistically significant. Yet looking only at Bennett's calculations—or, looking only at the data—one can never "know" for sure if the fish was thinking or not.

With statistics embedded everywhere via machine learning and algorithms, more than ever we risk seeing a dead fish thinking. In most statistical research, researchers rely on an extensive array of qualifying phrases, where the uncertainty of the phrase "most likely explanation" supplants the sureness of the word "obviously"; out in the world—as we encounter statistics in governmental decisions and search results—we rarely get such nuance. Perhaps, then, the challenges of our current situation, the slipperiness of our understandings, are not so far from those that arise in another story of a fish—the Book of Jonah—where that fish must have known something about the complications of human thought, or at the very least, the discomfort of having eaten one.

Jonah and the Whale

As told in the Bible, Jonah was ordered by God to travel to the city of Nineveh and warn the residents of the consequences of their sins, urging them to repent. Rather than obey and do God's bidding, Jonah fled from God aboard a ship, which was soon beset by a tempest that endangered him and his pagan shipmates. In an effort to know whose god had been angered and caused such a storm, the ship's crew resorted to cleromancy: "'Come, and let us cast lots, that we may know for whose cause this evil is upon us.' So they cast lots, and the lot fell upon Jonah."[15]

By this particular falling of the lot (analogous to flipping a coin or rolling dice in modern times), the God of the Hebrews had spoken to Jonah and the boat's crew through chance, an enunciation of deistic will. Jonah urged his shipmates to throw him overboard. Soon after they did so, Jonah was ingested by a great fish, spending three days and three nights inside its belly. Here, chance, instead of nullifying the results of an experiment, proved its opposite: the cause of the storm and the will of God. In modern experimentation, if the results are due to chance, they tell us nothing, but, for Jonah, chance directly speaks of the truth of the world.

Here, we see how the operations of rational, empirical methods of contemporary statistics are not so far afield from millennias-old theological theories of chance, even if the mode of interpretation has radically changed. With his experiment, Bennett drew a hundred thousand lots inside the dead fish's head to see if it can empathize and take part in the world. And by random chance, this dead fish declared to the scientist that it partakes of the world.

Statisticians would like to be able to set the threshold of calculated significance just high enough to silence random fluctuations, to mute the god who would speak to the world through cleromancy. Yet at the same time, the threshold must not be set too high, for nature would instead become silent, unable to emerge from the

---

15 Jonah 1:7, New American Standard Bible.

depths of statistical analysis to reveal its secrets. Statistically determined science skirts a razor's edge between hearing truth in random fluctuations and ignoring the truth of an intelligible, measurable nature whose forces advance like clockwork.[16]

It took until 1710 for God to cease speaking through chance, a historical moment when randomness was instead set in opposition to God. In this year, a Scottish physician named John Arbuthnot published a short note, "An Argument for Divine Providence, taken from the Constant Regularity Observ'd in the Births of Both Sexes."[17] Using data from christenings in London from the period 1629–1710, a period in which male births outnumbered female births every year, Arbuthnot calculated the possibility of observing these results by chance. If the ratio of male to female births was even, the odds of seeing another eighty-two years of consistently more male births would be of vanishingly small probability: one in nearly five septillion (one half multiplied eighty-two times).[18]

For Arbuthnot, this infinitesimal chance provided evidence not merely of a difference in birthrate, but also for the existence of divine providence. If births were not random distributions of male and female sex, he concluded that there must be some other causal force at work, which for him could only be God's action in the world. Arbuthnot's "Argument" is one of the first instances of statistically testing a hypothesis: instead of God speaking through chance, for Arbuthnot, statistics allowed God to speak only through the *absence* of chance.

Thus, in contrast to the long-held belief that chance articulates some truth, the advent of hypothesis testing means its ability to

---

16 In hypothesis testing, these are commonly called type I (false positive) and type II (false negative) errors.

17 John Arbuthnot, "An Argument for Divine Providence, Taken from the Constant Regularity Observ'd in the Births of Both Sexes," *Philosophical Transactions (1683–1775)* 27 (1710): 186–190.

18 Stephen M. Stigler, *The History of Statistics: The Measurement of Uncertainty before 1900* (Cambridge, MA: Harvard University Press, 1986), 225–226.

speak that truth now appears in inverted form. On the one hand, if random chance can produce the same result that one observes—such as a glowing salmon brain—one must maintain the possibility that what was witnessed offers no necessary insights about the nature of the world. On the other hand, if one can calculate that what was witnessed was highly unlikely by chance alone, such as some event having a one-in-a-thousand chance of happening randomly, one would strengthen their belief that they are, in fact, witnessing nature at work. When cleromancy repeats what human instruments say, researchers must reject their initial hypothesis and move on to another experiment or theory.

To return to Jonah, even beyond the facts of fish and lots, the prophet runs up against one of the central questions that plagues statistical, and thus algorithmic, knowledge: How are researchers and the public to use these insights? What can be done as a result of the knowledge gained? After praying for God's intervention, Jonah is expelled from the fish and ordered by God, again, to go to Nineveh. "Then Jonah began to go through the city one day's walk; and he cried out and said, 'Yet forty days and Nineveh will be overthrown.' Then the people of Nineveh believed in God; and they called a fast and put on sackcloth from the greatest to the least of them."[19] The people of Nineveh believe God. They repent, fast, and dress in sackcloth. Even the King of Nineveh sits in ashes and fasts. Despite his trust in chance, Jonah grows angry when God recants his promise to destroy Nineveh.[20] Appearing perhaps as more pagan than prophet, Jonah is upset by God's clemency. After Jonah's whole gastric ordeal, he would fail to see that what he was told was prophecy: a Nineveh destroyed.

---

19 Jonah 3:4–5, New American Standard Bible.

20 Not only for Jonah is the question of lots and chance in Judeo-Christian religion problematic, but throughout the Hebrew Bible, contradictory statements are given on the appropriateness of drawing lots. Take, for example, Proverbs 16:33: "The lot is cast into the lap, but its every decision is from God"; as contrasted by Leviticus 19:26: "Neither shall you practice enchantment, nor observe times" (usually taken to be a proclamation against divination or casting lots).

Jonah appears today as a ridiculous, obstinate figure. Unlike Cassandra, who was cursed to have her prophetic knowledge of the fall of Troy ignored, Jonah was believed and the people of Nineveh repented. Yet, he would rather be right than believed. He preferred a stubborn, unwavering God who foretells and enacts a single future, rather than one who accepts penance, forgives and opens the possibility of a different future. If Jonah were spewed from the fish's mouth in the age of big data, he would likely abandon his faith and convert to the god of predictive algorithms and the iron laws of history. Renouncing the subject of voluntaristic revolutionary theory, Jonah would prefer the side of the philosophers from Marx's famous quote: "Philosophers have only interpreted the world, in various ways. The point, however, is to change it."[21]

Despite trust in chance—or its inversion, in the case of hypothesis testing, where chance leads to the rejection of a hypothesis—we have found ourselves in the same position as Jonah: Does a crisis averted speak to the power of prediction, or does it negate that very power? Do predictive algorithms' success, then, speak to the impossibility of reconfiguring our political-social systems, or, on the contrary, do they provide the means of some as-yet unrealized promise of Demeter, who Antipater, quoted in the epigraph to this part, believed would free humanity from labor and the injustices it appears to demand?[22] Ultimately, the metaphysics and interpretations we provide regarding probability, and the statistics and machine learning algorithms that grow out of it, determine what humanity is capable of doing with this knowledge.

Let us return, then, to our dead salmon, who we hope, for the sake of reason, remains really and truly dead. Clearly, it is ridiculous to believe that a dead fish could think, let alone empathize. While we now know what went wrong in the statistical method—the incredible

---

21 Karl Marx, "Theses on Feuerbach," in *Karl Marx: Selected Writings*, Lawrence Simon, ed. (Indianapolis: Hackett Publishing, 1994).

22 Alfred Sohn-Rethel, *Intellectual and Manual Labour: A Critique of Epistemology* (Atlantic Highlands, NJ: Humanities Press, 1978), 2.

number of calculations done for analysis proportionally increased the risk of a false correlation—scientists still repeat this multiple-testing failure, and similar errors, ad nauseam in modern academic science. The slipperiness of these concepts, and the openness of probabilistic statements to the infinitesimally small possibility that an observation is due to chance alone, opens a space where a whole host of economic and material realities can intervene in these processes.

Outside academic laboratories and fMRI machines, but inside the everyday world of predictive algorithms, the instability of these concepts is continually at work. As probability (and chance) slides from the mark of God's presence to the guarantee of his absence, it leaves humans to confront its aporetic power and mystery, its ability to simultaneously reveal great truths and leave us groundless and uncertain.[23] Probability allows humanity to understand an uncertain world but also leaves it exposed to the metaphysical implications of this uncertainty.

The mysteries of algorithms, statistics, chance and inference are all vexing problems. But they remain problems whose logic and operation is central to our current situation. To fully pull apart these questions around chance and the production of knowledge requires that we trace the development of the meaning of probability in modern statistics. We must account for both the metaphysical foundations of statistics and the political economy in which it functions—two aspects that are intimately tied together. To do so will provide insight into both the possibilities and challenges of statistically produced knowledge, especially as it is given over to the demands of modern capitalism.

---

[23] Gerd Gigerenzer and Julian Marewski argue that statistics and scientific inference have continually sought an "idol of a universal method for scientific inference" and, failing that, have continued to produce surrogates to stand in for this absent idol, which even when they are productive in limited uses, continue to fail in their universal pretensions. Gerd Gigerenzer and Julian N. Marewski, "Surrogate Science: The Idol of a Universal Method for Scientific Inference," *Journal of Management* 41, No. 2 (2015): 421–440.

Chapter 5
Induction, Behavior and the Fractured Edifice of Frequentism

While employed as a statistician performing agricultural research at the Rothamsted Experimental Station in Eastern England, Ronald Fisher developed a system for experimental design that provides rigorous methods for testing a hypothesis. In important respects, it mirrored the intuition of John Arbuthnot, discussed in the previous chapter, that the inability of chance alone to explain some outcome provides proof that some other mechanism is at work. Out of this work Fisher published two influential books in the burgeoning field of statistical analyses: *Statistical Methods for Research Workers* in 1925 and *The Design of Experiments* in 1935. In these two texts, Fisher aimed to aid researchers in addressing the central challenge of a significant portion of modern scientific research: how to determine whether observed difference is due to experimental conditions rather than mere chance.

In *The Design of Experiments*, Fisher presents us with the case of Dr. Muriel Bristol and a cup of tea with milk, an experiment Fisher calls "the lady tasting tea." Bristol, a colleague of Fisher's at Rothamsted in the 1920s, had boldly boasted of her ability to discern through taste whether tea or milk is first added to a cup.

Within the office, the question of verification arose, a more complicated question than it may seem at first glance. If one offered Bristol a cup of tea with milk and asked whether tea or milk was added first, she could simply guess and have a 50 percent chance of that guess being correct; so even a correct identification is not proof of her ability. This is precisely the challenge that statistical testing attempts to address. So, in order to substantiate her claims, Fisher designed an experiment.

Fisher proposed the following solution: prepare eight cups of tea, four with the milk added first and four with the tea added first. Randomize their order, then ask Bristol to identify which four were prepared with milk first, providing her the ability to compare the cups with each other. Fisher calculated that there were seventy unique permutations of correct and incorrect guesses—so the odds of guessing all correctly are one in seventy, or approximately 1.4 percent. Conversely, there are seventeen possible permutations to guess three or better correctly out of the four with milk, which gives a nearly 25 percent chance of guessing correctly. With these odds in mind, we would likely accept that Bristol is truly able to determine how the tea is prepared if she selects all four cups correctly. While Fisher does not report the outcome of the experiment, according to secondhand sources Bristol correctly selected every cup of tea.[1]

We can also imagine a scenario where Bristol's skill only allows her to guess correctly 90 percent of the time—or even just slightly better than random guessing—that makes detection of her ability significantly more complicated. In such a case, Fisher would need to increase the number of cups in order to detect a partial ability to identify the preparation.[2] Yet while Fisher could serve Bristol thousands of cups of tea in order to decrease the possibility of chance, he could never fully eliminate it. Even if the odds become

---

1 David Salsburg, *The Lady Tasting Tea: How Statistics Revolutionized Science in the Twentieth Century* (New York: Henry Holt, 2002).

2 Kevin R. Murphy, Brett Myors and Allen Wolach, *Statistical Power Analysis: A Simple and General Model for Traditional and Modern Hypothesis Tests*, 4th ed. (New York: Routledge, 2014).

one in 1 million, there always remains the lingering 0.000001 percent, the remaining signifier of chance's immortality: the perennial presence of probability.

The P-value

While chance never dies, statistical methods must nonetheless declare it practically so in order to make any worthwhile claim about the world. Here, we arrive at the well-known concept of the "p-value," or probability value.[3] In the case of the lady tasting tea, the p-value would be presented as 0.014—if Bristol were actually guessing at random, there is still a 1.4 percent chance that she would guess all the cups correctly. While calculations of this value date back to centuries-old work by statisticians like Pierre-Simon Laplace and Karl Pearson, it was ultimately Fisher who popularized its calculation and highlighted its importance for modern hypothesis testing. The p-value has, in part on Fisher's suggestion, become one of the primary indicators of scientific validity. Fisher famously recommended that any result with a p-value above 5 percent be considered nonsignificant, or conceivably due to chance:

> It is usual and convenient for experimenters to take 5 percent as a standard level of significance, in the sense that they are prepared to ignore all results which fail to reach this standard, and, by this means, to eliminate from further discussion the greater part of the fluctuations which chance causes have introduced into their experimental results. No such selection can eliminate the whole of the possible effects of chance.[4]

While Fisher's negative claim was that any p-value calculation above 0.05 should be considered nonsignificant, many fields,

---

[3] Ronald L. Wasserstein and Nicole A. Lazar, "The ASA's Statement on P-values: Context, Process, and Purpose," *American Statistician* 70, No. 2 (2016): 129–133.
[4] Ronald Fisher, *The Design of Experiments* (New York: Hafner Press, 1971).

from psychology to agriculture, have interpreted it to mean its positive inverse: any result below 0.05 is to be considered significant. This difference may seem trivial and semantic—and many scientists treat it as such—but, it is hugely consequential. One science reporter commenting on the misuse of p-values in hypothesis testing went so far as to state: "Statistical techniques for testing hypotheses... have more flaws than Facebook's privacy policies."[5]

The misuse of p-values as the guarantor of statistical significance, and thus a means of understanding the results of scientific experiments, has become so problematic that the American Statistical Agency (ASA) went so far as to explicitly release "a formal statement clarifying several widely agreed upon principles underlying the proper use and interpretation of the p-value" in order to "improve the conduct or interpretation of quantitative science, according to widespread consensus in the statistical community."[6] While the ASA was rather measured in its tone, the statement was a clear and forceful rejoinder by the statistical community to the broader scientific community.

Fisher's original intention was for any p-value below 0.05 to serve only as a call for further research, not as the establishment of scientific fact. But too often now, such a result is translated as publishable proof. While there is a long history that has brought us from Fisher to our present moment—some of which is traced below—one of the major factors contributing to this shift has been the desire to make statistics into an easily followable set of steps for determining scientific truth—in essence to make it automatable, rather than the interpretative tool the founders of the field (and many present statisticians) envisioned.[7]

---

5 Tom Siegfried, "To Make Science Better, Watch Out for Statistical Flaws," *Science News Context Blog*, February 7, 2014, cited in Wasserstein and Lazar, "ASA's Statement on P-values."

6 Wasserstein and Lazar, "ASA's Statement on P-values," 131.

7 Jonathan Sterne and George Davey Smith provide a good short history of p-values and their use in the medical field, as well as an accessible technical

In many ways, modern statistics has been a victim of its own success. Statistical analysis's ability to evaluate diverse types of data has supplied the epistemic grounding to construct entirely new cottage industries, such as its oft-celebrated use to forecast elections using multivariate, aggregated datasets, or even to predict civil wars by examining countries' macroeconomic indicators alongside semantic analyses of domestic journalism.[8] But, as the growth of cheap, accessible computing power and availability of gigantic datasets continues to expand, statistical results with low p-values are still used as the positive establishment of correlation-turned-fact, rejecting the need for critical reflection. A correlative output that is right most of the time gets treated as truth, not as a provisional mathematical output based on a selected set of data.

Rather than engage in the complex philosophical debates that underwrite statistics, most statistical applications have ignored any underlying conflicts, banking instead on the widespread acceptance of these methods by everyone from colleagues to journal editors to science reporters.[9] Fisher was much more guarded. In setting up the "lady tasting tea" experiment, he recommended something that was then novel: the creation of a null hypothesis, or a general position declaring that there exists no relationship between two measured phenomena, or that two populations are the same in regard to the measured variable (e.g., the treatment and control group both had similar survival rates). A null hypothesis, then, is what an experiment is designed to find evidence against, and a successful experiment is one that denies the null hypothesis.

Yet, contrary to what is usually taught in contemporary introductory statistics classes, Fisher does not propose the creation of an "alternative hypothesis": the functional opposite of the null

---

explanation of some of the issues with their use: "Sifting the Evidence: What's Wrong with Significance Tests?," *Physical Therapy* 81, No. 8 (2001): 1464–1469.

8 "Methodology," Predictive Heuristics official website.

9 Wasserstein and Lazar, "ASA's Statement on P-values."

hypothesis that would describe the success of the experiment (in Hegelian terms, the negation of the negation). Fisher states this categorically: "Every experiment may be said to exist only in order to give the facts a chance of disproving the null hypothesis."[10] One could find evidence of a theory, such as the existence of the ability to distinguish the order of tea preparation, but for Fisher this is an ineligible hypothesis because it is ambiguous.[11] In Fisher's design of experiments, one simply gathers evidence against the possibility that chance alone accounts for what was observed.

In philosophical terms, Fisher's theory of statistical inference is ultimately a theory of difference. The base question that undergirds statistical work is whether or not two groups are different. In the case of medicine, a researcher may seek to know whether or not a new treatment improves outcomes and so tries to detect a difference between a control group (given a placebo) and a treatment group. In the case of the "lady tasting tea," Fisher wanted to know if Bristol's guesses as to whether milk or tea was added first differs from random chance. Thus, the experimental difference offers evidence for her ability to detect difference.

We might trace this fundamental concern with difference back to Fisher's early agriculture work, where he and others labored over the determination of difference between various cultivation methods, different plant species and different climates. In this work, everywhere one looks, one sees difference. Consequently, research for Fisher was merely the ability to disprove the null hypothesis that no difference existed. In such a framework, the primary role of science is not to understand causal mechanisms but rather to ascertain when two groups differ. For Fisher, the universal law that can be derived from data is the existence of difference.

---

10 Fisher, *Design of Experiments*, 16.
11 Fisher does suggest that one could formulate a hypothesis on this latter case, but warns that "this hypothesis could be disproved by a single failure, but could never be proved by any finite amount of experimentation." Ibid., 16.

Fisher, Race and Difference

Following his 1925 *Statistical Methods*, Fisher's next book, *The Genetical Theory of Natural Selection*, was published in 1930 and served as one of the first attempts to combine the study of genetics with Darwinian evolution.[12] Fisher's advancement of this work was deeply racist. Soon after its publication, Fisher left the agricultural world of the Rothamsted Experimental Station and began his first academic appointment in the Eugenics Department at University College London. Fisher spends the entire last third of his *Genetical Theory* developing his theory of eugenics, particularly the then-nascent field of "population genetics" that dealt with ideas like the "decay of civilization" resulting from declining birth rates among the upper classes.[13] Even more damning is that he demonstrated a continued commitment to his eugenic theories even after World War II, when many other scientists retreated from the promotion of eugenic science.

In 1950, the United Nations Educational, Scientific and Cultural Organization (UNESCO) convened a group of scientists to draft a statement on what they called "the nature of race," intending the statement to direct scientific and other endeavors away from various racist positions. In this statement, the concept of racial difference was declared to be bankrupt on the grounds that there was a lack of scientific evidence for this difference and that even if such difference could be proven scientifically in the future, it would be morally irrelevant for society and politics:

> At the present moment, it is impossible to demonstrate that there exist between "races" differences of intelligence and temperament other than those produced by cultural environment. If, tomorrow, more accurate tests or more thorough

---

12 Ronald Fisher, *The Genetical Theory of Natural Selection* (Oxford: Oxford University Press, 1999).
13 Ibid.

studies should prove that "races" as such do, in fact, have different innate faculties or aptitudes, UNESCO's moral position on the race question would not be changed.[14]

Fisher was party to these discussions, but ultimately ended up objecting to their conclusion. A commentary on the statement includes a summary and quotations from Fisher's position, which stress his commitment to theories of racial difference.[15]

Fisher's racism, and his objection to the UNESCO statement, is best understood through his unwavering belief in difference. In his statistical terms, Fisher rejected the null hypothesis that races are randomly selected from the population. With these statements, we see further proof of the extent to which his ideology and conception of the scientific method was dedicated to the detection of difference, especially in relation to populations and genetics. He was, quite literally, a man who believed in the truth of blood and soil: an agriculturalist and a scientific racist.

This connection between Fisher's statistics and his racism points squarely to the sociopolitical importance of the metaphysical foundations of statistics. This is not to say that all statistics, or even the use of Fisherian methods, are necessarily racist. Rather, it is possible to see in Fisher's metaphysics a tendency that ultimately shaped his politics: his commitment to difference caused him to see difference as the underlying truth of the world. Moreover, his understanding of statistics as a science provided the means to translate his racism into objectified form, giving it (from his perspective) the form and force of truth. Just as the commodity form of exchange predetermines the possible forms of economic knowledge, the grounds of this probabilistic knowledge shape the very political and social possibilities—the conditions of what and how it is possible to know—of our statistically (and, increasingly, algorithmically) mediated world. His belief in the truth of

---

14 UNESCO, *The Race Question*, July 18, 1950, 3.
15 Ibid., 27.

difference provided fertile ground for his racialized view of the world and offered it the force and language of science.

Alongside these ideological shortcomings, Fisher's conception of science was also a very methodologically conservative one. Fisher advanced a theory of scientific induction that amounts to an inverse form of the Austrian-British philosopher Karl Popper's famous formulation of falsifiability. For Popper, scientific theories are never provable, but are rather falsifiable—experimentation can only prove them false.[16] But Fisher's theory is even more radical and skeptical of the inductive power of science. For him, one cannot even falsify a theory. One can only falsify the *absence* of a theory, namely the null hypothesis, with the only possible progress in science being an increasing belief in the existence of difference. If one finds substantial evidence against a null hypothesis, it provides no guarantee of a specific theory.

It was in reaction to this conservatism in Fisher's approach that other statisticians, like Jerzy Neyman and Egon Pearson, attempted to build upon Fisher's work. At first, Fisher was interested in their efforts. But soon he responded with vitriol, describing their work at one point as "childish" for its disagreements with his approach.[17] Neyman and Pearson responded to Fisher with equal intensity, leading to a conflict that was both deeply philosophical and petty, which carried on for decades, depositing evidence of hostility across the scholarly record.

---

16 Karl Popper, *The Logic of Scientific Discovery* (New York: Routledge, 1992). For a discussion of the relationship between Fisher's scientific induction and Popper's account of scientific falsifiability, particularly in reference to psychology, see Paul E. Meehl, "Theoretical Risks and Tabular Asterisks: Sir Karl, Sir Ronald, and the Slow Progress of Soft Psychology," *Journal of Consulting and Clinical Psychology* 46, No. 4 (1978): 806–834.

17 Gerd Gigerenzer, Zeno Swijtink and Lorraine Daston, *The Empire of Chance: How Probability Changed Science and Everyday Life* (Cambridge, UK: Cambridge University Press, 1990), 105.

Neyman and Pearson

While one can trace the specifics of this conflict through various books and articles, three main issues underlay Neyman and Pearson's claims against Fisher. First, they argued that Fisher's method was only able to handle false positives (what they call "errors of the first kind") and lacked any way to quantify "errors of the second kind," or false negatives—that is incorrectly believing that what is observed is merely due to chance, or in terms of hypothesis testing, the incorrect rejection of a hypothesis (now known as "type II errors.").[18] While Fisher's method uses a p-value to represent the probability that there is no real difference, any researcher employing his method lacks the means to quantify the possibility that there was a difference that went undetected (a false negative).

Second, Neyman and Pearson took issue with Fisher's interpretation of a statistical test's result. In Fisher's inverted Popperism, a statistician reaches a point—often the p-value cutoff of 5 percent—at which they can hold a reasonable belief that difference, in fact, exists. The problem with this "point," Neyman and Parson argue, is that if one accepts a cutoff value of 5 percent, as has since become common in many fields, then 5 percent of all tests will see the null hypothesis incorrectly rejected.[19] Returning to the fMRI problem from the previous chapter, we can now better understand the importance of a critical evaluation of such thresholds: there, given the number of tests, a 5 percent cutoff is way too high and leads to spurious results; but conversely, setting it too low risks throwing out important results.

Some have suggested the threshold simply be lowered. However, as statistician Andrew Gelman has recently argued,

---

18 Ibid., 99.
19 Demetrios N. Kyriacou, "The Enduring Evolution of the P Value," *Jama* 315, No. 11 (2016): 1113–1115; David Jean Biau, Brigitte M. Jolles and Raphaël Porcher, "P Value and the Theory of Hypothesis Testing: An Explanation for New Researchers," *Clinical Orthopaedics and Related Research* 468, No. 3 (2010): 885–892.

lowering this 5 percent threshold is not enough to restore confidence in the results of statistical analysis of experiments.[20] Instead, Gelman recommends statisticians embrace the uncertainty of statistical testing by acknowledging that results are contingent upon the decisions scholars make when they formulate hypotheses and select, clean, and interpret data—steps that are crucial to any analysis that aims to reveal relationships between variables and hypotheses.[21]

Third, as a corollary, Neyman and Pearson point to a deeper philosophical issue at work in one of the ways frequentism is often understood—one they felt did not deal well enough with the fact that it is nonsensical to make probabilistic statements about single, concrete events. In the problem of the "lady drinking tea," if Fisher's statistical test suggests after an experiment that there is a 5 percent chance that Dr. Bristol guessed randomly, this is nonsensical: either she guessed randomly or did not. Likewise, either it rains tomorrow or it does not—there is nothing properly probabilistic about it. Especially if the result of an experiment is supposed to measure something that concretely exists in the world, it is in general a challenge to integrate probabilistic statements into any scientific epistemology. This is a major problem that lies at the heart of statistics: How is it possible to provide solid ground to knowledge that is derived from probabilistic statements? Or, to phrase it in its converse, how is it possible to speak probabilistically about a concrete event, even if we do not know whether or not it happened?

To deal with this challenge, Neyman and Pearson attempted to provide a stronger interpretation of the results of tests, a solution Neyman termed the principle of "inductive behavior."[22] This principle proposes that over a long series of statistical tests, a

---

[20] Andrew Gelman, "The Problems with P-values Are Not Just with P-Values," *American Statistician* 70, No. 2 (2016): S1–S2.

[21] Ibid.

[22] Jerzy Neyman, "'Inductive Behavior' as a Basic Concept of Philosophy of Science," *Revue de l'Institut International de Statistique* 25 (1957): 7–22.

statistician will necessarily be wrong a given and computable number of times. So, one ceases to claim that the result of a study is necessarily true but rather behaves *as though* it is true; in this way, inductive behavior sidesteps the fact that frequentism demurs from describing single events and allows researchers to act as if it could.[23]

To mitigate the negative effects of occasionally being wrong, one sets the threshold for the rejection of a hypothesis not at some anointed quantity—such as 5 percent—but at a calculated level that takes into account the costs of each possible outcome. Whereas Fisher's experiments only tested a null hypothesis, Neyman and Pearson recommended the construction of two alternative hypothesis, between which a statistical test could select. With this approach, they reasoned, one could calculate an optimal behavior based on the probability of various outcomes over multiple tests, as well as the economic costs of being wrong in either direction. For example, an algorithm that prescreens job applications would require more human work if it let in unqualified candidates (a false positive, or type I error) but would likely have even-greater negative consequences if it excluded a well-qualified candidate from the pool (a false negative, or type II error). One could quantify the economic costs in either direction and determine the economically ideal sensitivity.

This economic framing is, then, why in the Neyman-Pearson approach it is so important to recognize errors of the second kind. In any given statistical test, there is a risk and an accompanying material cost associated with both a false positive (i.e., the incorrect assumption that there is difference) and a false negative (i.e., the incorrect assumption that there is no difference). Like a profitable casino, if the costs and risk are appropriately calculated, the

---

23 Hacking argues that inductive behavior may still be a form of "inference." See "The Theory of Probable Inference," in *Science, Belief, and Behaviour: Essays in Honor of R.B. Braithwaite*, D.H. Mellor, ed. (Cambridge, UK: Cambridge University Press, 1980), 141–160.

house is bound to lose some hands, but over time the house will ultimately win more than it loses. Similarly, a researcher can calculate an experiment's ability to detect an effect, balancing these costs and benefits prior to the study. For example, as one increases an experiment's sample size, it may become easier to detect an effect, minimize the impact of randomness, and hence avoid errors of both the first and second kind. But at the same time, the larger the sample size, the more expensive the experiment. Thus, even if one is ultimately wrong some of the time, in the long run one will end up making the most profitable decisions.

A recent reversal by the United States Preventive Services Task Force (USPSTF) of one of their recommendations demonstrates the appeal of this behavioral approach. In 2011, the USPSTF recommended that men not receive the prostate-specific antigen (PSA) blood test, a test designed to detect early signs of prostate cancer.[24] While the PSA test can discern life-threatening health ailments, it also is encumbered by a high rate of false positives (type I errors), with some estimates suggesting 75 percent of patients who had elevated PSA did not actually have cancer.[25] Furthermore, health researchers discovered that even for men who had tumors on their prostate, many of those tumors would grow so slowly that they would not be of vital concern during one's lifetime. The cost of false positives in PSA tests has been high: ensuing treatment involves risks of incontinence, erectile dysfunction, bowel complications and infection. And after weighing these costs and the low benefits to early treatment, the USPSTF recommended against the test.

But in 2017, the USPSTF relaxed their recommendation against PSA testing.[26] This updated suggestion did not come as a result of

---

24 Roger Chou, Jennifer M. Croswell, Tracy Dana et al., "Screening for Prostate Cancer: A Review of the Evidence for the U.S. Preventive Services Task Force," *Annals of Internal Medicine* 155, No. 11 (2011): 762–771.

25 "Prostate-Specific Antigen (PSA) Test," National Cancer Institute official website.

26 In their grading system, they went from a "D" to a "C"; in short, from opposition to a neutral position.

a more effective test or a less dangerous treatment. Rather, what had changed between 2011 and 2017 was doctors' awareness of the risks of treatment. After the initial recommendation by USPSTF, doctors began sending fewer patients for surgical interventions and instead recommended further monitoring. This increase in caution decreased the cost of false positives and was ultimately enough to change the recommendation.

This behavioral approach, which attempts to assign calculable value to false negatives and positives, is especially well suited for business, medical and other types of decision making. In these contexts, one can quantify the material costs of false negatives, false positives, and additional testing. But when applied to the general production of scientific knowledge, it becomes far more complicated to quantify the costs of various types of error. Because of this difficulty, Fisher attacked Neyman and Pearson relentlessly, stating, "To do so would imply that the purposes to which new knowledge was to be put were known and capable of evaluation... As workers in Science we aim, in fact, at methods of inference which shall be equally convincing to all freely reasoning minds, entirely independently of any intentions that might be furthered by utilizing the knowledge inferred."[27] Elsewhere Fisher states, in regard to the essence of scientific discoveries, that he does "not assume that they are capable of evaluation in any sort of currency."[28]

A number of commentators have noted how appropriate Neyman and Pearson's approach is to industrial applications, like quality control systems that evaluate products by testing samples against a standard specification, where it is more important to be right *often enough* than it is to always be right.[29] These differences

---

27 Ronald Fisher, *Statistical Methods and Scientific Inference*, 3rd ed. (London: Collins Macmillan, 1973), 106–107.

28 Ronald Fisher, "Statistical Methods and Scientific Induction," *Journal of the Royal Statistical Society*, Series B, *Methodological* 17, No. 1 (1955): 77.

29 See ibid., 69–78; and Stephen Spielman, "A Refutation of the Neyman-Pearson Theory of Testing," *British Journal for the Philosophy of Science* 24, No. 3 (1973): 201–222.

in the ultimate meaning of statistical tests did not escape Fisher, who criticized Neyman and Pearson's approach for its technological and managerial focus at the expense of what he believed was "pure science" and holistic knowledge. Fisher explicitly framed this difference in relation to the politics of the Cold War:

> I shall hope to bring out some of the logical differences more distinctly, but there is also, I fancy, in the background an ideological difference. Russians are made familiar with the ideal that research in pure science can and should be geared to technological performance, in the comprehensive organized effort of a five-year plan for the nation. How far, within such a system, personal and individual inferences from observed facts are permissible we do not know, but it may be safer, and even, in such a political atmosphere, more agreeable, to regard one's scientific work simply as a contributory element in a great machine, and to conceal rather than to advertise the selfish and perhaps heretical aim of understanding for oneself the scientific situation. In the U.S. also the great importance of organized technology has I think made it easy to confuse the process appropriate for drawing correct conclusions, with those aimed rather at, let us say, speeding production, or saving money. There is therefore something to be gained by at least being able to think of our scientific problems in a language distinct from that of technological efficiency.[30]

If Fisher was practically and ideologically an agriculturalist, Neyman and Pearson were wholesale industrialists. Their interpretation of statistical inference abandons truth in favor of economic efficiency, marking a stark turn toward neoliberal thinking in the sciences. This economic interpretation replaces the aim of scientific knowing as an end in itself with a knowledge subsumed in the calculation of economic gain. Where Fisher tested a null

---

30 Fisher, "Statistical Methods and Scientific Induction," 70.

hypothesis, Neyman and Pearson set up alternatives, assigning costs for erring in either direction in order to maximize profit in the long run. And so, while Neyman and Pearson were able to construct a philosophically rigorous interpretation of statistical testing, challenging Fisher's established statistical epistemology in the process, the efficacy of this new interpretation came at the expense of undermining the solidity of any specific conclusion, which for them cannot be declared either fully true or false.

Leonard Savage, one of the founders of subjective theories of probability that we will explore later, puts the situation directly: "The traditional idea of inference as opposed to behavior seems to me to have its roots in the parallel distinction between opinion and value."[31] Contrary to Fisher's attempts to hone his own opinion of the truth, Neyman and Pearson sought to eschew opinion in favor of the production of value. For Neyman and Pearson, truth—under the reign of this "as if" of behavioral induction—becomes relative to its economic value and thus also tied to the shifting and relativized sands of the economy.

Hybridization

While ideological and methodological differences grew between Fisher and Neyman and Pearson, other researchers recognized the power of both statistical approaches, but failed to comprehend the theoretical differences between them. This recognition rapidly led to a haphazard hybridization of Fisher's framework for null hypothesis testing with Neyman and Pearson's alternative hypotheses—where one compares a null hypothesis to a single alternative hypothesis—as well as a simplistic distillation of these statistical

---

[31] Leonard Savage, "The Foundations of Statistics Reconsidered," *Proceedings of the Fourth Berkeley Symposium on Mathematical Statistics and Probability*, Vol. 1, *Contributions to the Theory of Statistics* (Berkeley: University of California Press, 1961).

methods into textbooks for fields (such as experimental psychology) that had neither the time nor energy to engage with the thorny philosophical problems at issue.[32]

In these texts, critical background—discussions of epistemological debates, the difference between the behavioral approach and Fisher's conservatism regarding the interpretation of results, and even the names of Fisher, Neyman and Pearson—was dropped. An uncritical hybridization of once-incommensurate systems created a chimeric monster—one that could be called the Fisher-Neyman-Pearson model. This monster is what many now know as "statistics," and what is currently taught to classes of undergraduate and graduate students across the world.

This hybrid system was highly effective and served a much-needed purpose, enabling post–World War II advances in fields ranging from biology to the social sciences. In these fields, such hybrid probability measures provided a means to evaluate results across an increasingly complex research landscape. When results were relatively straightforward, this model was capable of quantifying results and providing a cutoff (p-value less than 0.05) for publication. Still, this hybrid approach amounted to an incoherent amalgamation: Fisher's trust that statistical results moved ever closer to truth was combined with Neyman and Pearson's economically driven "close-enough" approach.

This methodological, and ultimately philosophical, hybridization of statistics has since established a scientific-academic complex that produces knowledge like a factory—experiment, publish, receive a grant, and repeat—but treats the product as though it were immune from the dictates of economy.[33] In this

---

[32] Gigerenzer and Switjink, *Empire of Chance*; Carl J. Huberty, "Historical Origins of Statistical Testing Practices: The Treatment of Fisher versus Neyman-Pearson Views in Textbooks," *Journal of Experimental Education* 61, No. 4 (1993): 317–333.

[33] Scholars writing on the effects of capitalism and the ideology of neoliberalism on the university have convincingly made similar claims. See Stanley Aronowitz, *The Knowledge Factory: Dismantling the Corporate University*

knowledge factory, there is no plant manager or five-year plan to designate a research program or mission; no one has calculated the costs of errors of various types. Instead, these costs are borne by a combination of the poor researchers whose careers disappear with their once-statistically significant results, by the patients whose lives are ruined by ineffective treatments or horrific side effects, and by all the other environmental and individual impacts of ineffective research practices.

This is not to say such a scientific-academic complex does not "work." In fact, it functionally works, again, like a casino: individuals may win or lose on any given day, but, on the whole, the house (research labs, corporate research and development) keeps winning. But as the sciences have grown more and more complex—in terms of their infrastructure, specialization, costs, requirements of prior knowledge, economic demands and so on—the house's margins keep shrinking, a fact made ever more dire since there is no one actually managing the house.

Here, it is helpful to describe one of the most significant changes that this hybrid statistics has wrought in information theoretic terms: statistics provides a set of tools for distinguishing signal from background noise—just like we saw in the case of fMRI analysis. The stronger the signal and the weaker the noise, the easier it is to detect. For example, in the case of the "lady tasting tea," it is much easier to detect her abilities if she is always correct (a strong signal) than if she is correct at a rate slightly better than she would on pure chance (a weak signal, especially considering the background noise that someone could randomly guess correctly half of the time).

But many of the very strong signals have already been discovered: electricity, penicillin and general relativity, among others. Scientific research now requires more and more energy to discover

---

and *Creating True Higher Learning* (Boston: Beacon Press, 2000); Henry Giroux, "Neoliberalism, Corporate Culture, and the Promise of Higher Education: The University as a Democratic Public Sphere," *Harvard Educational Review* 72, No. 4 (2002): 425–464; and Fred Moten and Stefano Harney, "The University and the Undercommons," *Social Text* 79, No. 2 (2004): 101–115.

weaker signals, and in many human sciences, this means very large sample sizes for experiments. This can be seen even more clearly in the development of particle physics. Maurice Goldhaber, former director of Brookhaven National Laboratory in New York, puts it poetically: "The first to disintegrate a nucleus was Rutherford, and there is a picture of him holding the apparatus in his lap. I then always remember the later picture when one of the famous cyclotrons was built at Berkeley, and all of the people were sitting in the lap of the cyclotron."[34] Now, scientists build particle accelerators the size of small cities.

The French philosopher Michel Serres concurs: "The first shepherd lays his hands on the treasure of the scrolls found in the cave; there are a hundred thousand. Now, with electronics and international relations, you glean rare, scattered, barely noticeable atoms of letters. Newton under the apple tree, all alone, gives the law of the world, leaving only a few marginal scraps for his innumerable offspring."[35] We might object that quantum theory and relativity are more than "marginal scraps," but it remains clear that more energy—and more precise instruments—are required to move knowledge into the rarified, scientific realms beyond Newtonian physics. Ultimately, the increased energy needed to detect smaller and smaller signals means that scientists' methods of differentiating signal from noise require much finer precision. Moreover, as the expense of research increases, the importance of political economy to science likewise grows.

This statistical hybrid may have worked marvelously for conventional science in its early post–World War II decades. But today, with much stronger demands for precision to detect weaker signals, its imprecision is undeniable. The sciences, notably social psychology and clinical medicine, are experiencing what

---

[34] Quoted in Steven Weinberg, "The Crisis of Big Science," *New York Review of Books*, May 10, 2012.

[35] Michel Serres, *The Parasite*, trans. Lawrence R. Schehr (Minneapolis: University of Minnesota Press, 2007).

contemporary commentators have termed "replication crises."[36] In this world of crisis, it appears likely that substantial amounts of scientific research are wrong, in large part because these sciences' methodological bases have provided the requisite wiggle room to furnish statistically significant results (often with a p-value just below 0.05) without actually proving anything beyond the fact that a study found statistical significance.

Thus, flexibility in research design and data analysis in the knowledge factory managed by this statistics has allowed researchers to fiddle with results just enough to make something academically publishable fall out of the data, a practice popularly known as "p-hacking."[37] For example, researchers occasionally exclude outliers, look at subgroups or add additional data—all practices that on their own do not constitute malpractice but in the aggregate can produce results that appear significant when they should not. Modern science may be finally running up against the limits of statistical methodologies. In response, some statisticians have argued that we should reconsider the standard models of frequentist science. Andrew Gelman, for one, writes: "If the classical theory of hypothesis testing had lived up to the promise it seemed to have in 1950 (fresh after solving important operations-research problems in the Second World War), then indeed maybe we could have stopped right there."[38]

More often than not, when attempts to replicate scientific studies fail—or errors are found in methodologies, such as with fMRI studies—the problem is not with the pure statistics, per se. Rather, as with fMRI, the issue is one of methods, software, data collection

---

36 John P.A. Ioannidis, "Why Most Published Research Findings Are False," *PLoS Medicine* 2, No. 8 (2005): E124.

37 Joseph P. SImmons, Leif D. Nelson and Uri Simonsohn, "False-Positive Psychology: Undisclosed Flexibility in Data Collection and Analysis Allows Presenting Anything as Significant," *Psychological Science* 22, No. 11 (2011): 1359-1366.

38 Andrew Gelman and Christian P. Robert, "'Not Only Defended but Also Applied": The Perceived Absurdity of Bayesian Inference," *American Statistician* 67, No. 1 (2013): 1-5.

and motivations. Indeed, much of contemporary scientific scholarship has found itself conditioned by professional and economic pressures: the need to publish, the need for journals to make money, funding from corporations, the immense drive to be respected by peers and so on. Stanford epidemiologist John Ioannidis—whose 2005 article "Why Most Published Research Findings Are False" caused deep reflection across scientific circles—argues that there must be a change in "the incentive and reward system in a way that would reward the best methods and practices. Currently we reward the wrong things: people who submit grant proposals and publish papers that make extravagant claims. That's not what science is about."[39]

Such a poorly designed statistical enterprise has ultimately created the grounds for a new anti-science: poor and distracting incentive structures, combined with researcher flexibility, add enough noise to the system to overwhelm many of the signals of meaningful discoveries.

The Noise in the Signal

There is another element of this "hybridization" in statistical thinking that is rarely recognized: in this scheme, a means of scientific inference based around an individual, knowing subject is transposed onto a collective system of knowing, such that what the individual researcher has worked out is taken to be known collectively. The vast majority of our philosophical approaches to statistics, save perhaps the strictest interpretation of Neyman and Pearson's work, assume that the one seeking knowledge is an individual. On the other hand, our systems of academic knowledge production treat these results as though they are collectively held. If anyone was aware of this difficulty it was Fisher, as evidenced

---

39 John Ioannidis, interviewed by Julia Belluz, "John Ioannidis Has Dedicated His Life to Quantifying How Science Is Broken," *Vox*, February 16, 2015.

above, who also contrasted the Soviet model of technologically driven science with the "selfish and perhaps heretical aim of understanding for oneself the scientific situation."[40]

Like the agrarian that he was, Fisher presented his system as built completely around the lone, self-sufficient researcher, or farmer who will take what they learn about their field and immediately implement it there. In contrast, the industrial Neyman-Pearson system required a bureaucrat to manage the production of knowledge, someone who can weigh the costs and benefits of various outcomes for the entire enterprise and not just their own career. Yet, while Fisher focused on the individual researcher, he was not unaware of the challenges that accompany the collectivization of knowledge. To this end, Fisher argued that negative results should be shared publicly so that researchers would have both positive and negative evidence to weigh.[41] If one only shares positive results, researchers do not have all of the evidence required to run the appropriate calculations and stock up evidence in the spirit of Fisher's inverse Popperism. Chance results will be published and acclaimed while countervailing evidence sits in dusty file drawers—an outcome that severely biases the scholarly record.[42]

Still, while Fisher would likely never have admitted it, even his "selfish" interpretation of statistics requires a manager or a bureaucrat for it to function properly. Without such oversight, there is no force or incentive to collect all of these failed tests. And in the current distributed model of research where there is no five-year plan or plant manager, there is very little value or potential in negative results. Negative results produce nothing productive for the contemporary scientific milieu; they generate neither money nor prestige. These negative results are rarely welcome in a world where surprising results can sell fame, patents, and journals. While

---

40 Fisher, "Statistical Methods and Scientific Induction," 70.

41 Gigerenze, Switjink and Daston, *Empire of Chance*, 108.

42 See Uri Simonsohn, Leif D. Nelson and Joseph P. Simmons, "P-Curve: A Key to the File-Drawer," *Journal of Experimental Psychology: General* 143, No. 2 (2014): 534.

there is an academic movement organizing to share negative results, the incentives of grant funding and tenure often mitigate the desire to actively do so.[43] The situation is even worse outside of academia, where companies like those in the pharmaceutical and tobacco industries can easily hide nonsignificant data and calculations, presenting curated, and thus manipulated, outcomes as eternal truths.[44]

Ultimately, the hybridized model creates what we have discussed, in earlier chapters, as a mystery: it produces a metaphysics that objectifies inequalities into equalities. It treats knowledge as value (i.e., based on a behavioral economics) while, as Fisher does, it claims its value cannot be evaluated—in other words, that research cannot and should not be managed with some larger plan. In short, statistics redoubles a sovereign objective perspective that equates individual knowing with collective knowing, and thus simultaneously brings this knowledge to market while claiming that it can still be rescued from the perversions and dissimulations of that market.[45] While processes of academic knowledge production use peer review and disciplinary conversation to collectivize knowledge, modern statistics functionally conflates individual and collective knowledge by applying Neyman and Pearson's philosophy of industrial production of knowledge while maintaining the veneer of Fisher's heroic individual scientist who is capable of knowing.

The Fisherian claim that statistical knowledge is truth (even in Fisher's conservative interpretation that claims only to detect

---

43 For example, *The Journal of Negative Results in BioMedicine*. Such efforts are clearly on the whole positive and should be supported, but it seems unlikely they will be able to save academic research from its crises on their own.

44 Lisa A. Bero, "Tobacco Industry Manipulation of Research," *Public Health Report* 120, No. 2 (2005): 200–208; Marcia Angell, "Industry-Sponsored Clinical Research: A Broken System," *Journal of the American Medical Association* 300, No. 9 (2008): 1069–1071.

45 Donna Haraway, "Situated Knowledges: The Science Question in Feminism and the Privilege of Partial Perspective," *Feminist Studies* 14, No. 3 (1988): 575–599.

difference) remains like the ghost of the Enlightenment subject, in the sense that its simultaneous presence and absence is what makes, and underwrites, the social and economic force of these calculations. While Neyman and Pearson see this statistical knowledge as contingent and temporary—placed under the cautious "as if" of their behavioral and economic approach—the patina of Fisher's claim makes it appear objective and true. To put it in Marxist vernacular, Fisher's work furnishes statistical knowledge the structure of the fetish: the result appears stripped of its context, able to think for itself, and is thus given objective force.

While this will not suffice to explain the totality of the phenomena, it is possible here to see an element of the excitement some portions of the public affix to scientific ideas and explanations that are largely discredited: from the return of scientific theories of racial and sexual difference to climate change denial to concerns that vaccines cause autism.[46] While far from Fisher's intent (and in fact nearly the exact opposite), his simultaneous belief in difference as the truth of science and the lone scientist's ability to know provide a dangerous combination that has empowered many scientists and nonscientists alike to believe that they alone are able to ascertain true difference, and often to fetishize the results of a single study (even a discredited one, in some cases).[47]

The p-value that researchers attempt to push below a certain threshold to prove the truth of an alternative hypothesis appears as a ghost of a departed quantity—of those present absences that mark the mysteries of mathematics and knowledge production. Yet philosophical and epistemological obfuscation guarantees that many researchers fail to grasp what is truly at stake, continuing to run study after study and test after test, while concerns about the viability of the larger enterprise mount.

---

46 Alondra Nelson, *The Social Life of DNA: Race, Reparations, and Reconciliation after the Genome* (Boston: Beacon Press, 2016).

47 Stephan Lewandowsky and Klaus Oberauer, "Motivated Rejection of Science," *Current Directions in Psychological Science* 25, No. 4 (2016): 217–222.

In this light, one should not be surprised that John Ioannidis concludes his short article on why most scientific research is false by writing, "For many current scientific fields, claimed research findings may often be simply accurate measures of the prevailing bias."[48] Here, again, we see the force of objectification, taking what amounts to preexisting bias (although here Ioannidis is referring to bias in favor of certain theories) and laundering it through some computation to give it the non-locatable force of truth. The hasty hybridization of these models of statistics has created a very profitable mathematical infrastructure, supporting a mutually reinforcing system that favors production over quality control. In this way, the conclusions that this giant research machine produces are not necessarily wrong, but the ground on which it operates is not nearly as stable as it is often made out to be.

This conceptual ground is founded upon a mystery—an equation of unequal elements that serves to objectify our belief in its function by treating value as truth. The instability of these mysteries—here on display in the philosophically contradictory hybridization of Fisher's science of truth with Neyman and Pearson's science of value—expose these metaphysical presumptions to the political economies within which they operate. It is both for this reason and as a result of it that statistics begins, with Neyman and Pearson, to try to find its ground in the act of producing value. Just like the commodity for Marx, the metaphysical magic trick of statistics both justifies and finds its truth in exchange; in the words of Alfred Sohn-Rethel, it is a real abstraction, producing abstract truths out of a material economy and claiming that they have no locus and thus are not social.

The point, then, is not to stop this machine of knowledge production, but rather to dig beneath its base. There, it may be possible to reconfigure its metaphysics, to lay new claims to this system's understanding of both knowledge and value. This is the work of a revolutionary mathematics. In short, to save this

---

48 Ioannidis, "Why Most Published Research Is False," 2005.

machinery from ruin requires an insistence that any crisis in the sciences today is first and foremost a crisis of capitalism. The only future the sciences have is one where they work against and beyond the economies that threaten its destruction.

PART III
Bayesian Dreams

If a universal mind existed, of the kind that projected itself into the scientific fancy of Laplace—a mind that could register simultaneously all the processes of nature and society, that could measure the dynamics of their motion, that could forecast the results of their inter-reactions—such a mind, of course, could a priori draw up a faultless and exhaustive economic plan, beginning with the number of acres of wheat down to the last button for a vest. The bureaucracy often imagines that just such a mind is at its disposal; that is why it so easily frees itself from the control of the market and of Soviet democracy. But, in reality, the bureaucracy errs frightfully in its estimate of its spiritual resources. In its projections it is necessarily obliged, in actual performance, to depend upon the proportions (and with equal justice one may say the disproportions) it has inherited from capitalist Russia, upon the data of the economic structure of contemporary capitalist nations, and finally upon the experience of successes and mistakes of the Soviet economy itself. But even the most correct combination of all these elements will allow only a most imperfect framework of a plan, not more.

—Leon Trotsky, "The Soviet Economy in Danger"

Chapter 6
Bayesian Statistics and the Problem with Frequentism

Frequentism may have worked well in the past, but today it struggles. Many now look elsewhere for statistical insight, often by turning to Bayesian statistics. Despite being named after the eighteenth-century mathematician and reverend Thomas Bayes, these methods have only entered the mainstream in the last few decades. In fact, Bayes did not invent the theorem that bears his name, which serves as the mathematical foundation of Bayesian statistics—a feat that was accomplished by astronomer and polymath Pierre-Simon Laplace in the early nineteenth century.[1] But, in a posthumously published essay, Bayes laid out the basic math of this approach, which allows for the estimation of some unknown value from data (such as the estimation, after observing ten lottery

1 Stephen Stigler suggests that it may have been possible Nicholas Saunderson discovered the basic principles of Bayesian approaches prior to Thomas Bayes and informed David Hartley, who makes brief mention of this approach in his 1749 *Observations of Man*. Stigler provides an insightful and amusing weighing of the evidence and in the end uses Bayesian analysis to calculate the probability of whether Bayes or Saunderson initially made the discovery. "Who Discovered Bayes's Theorem," in *Statistics on the Table* (Cambridge, MA: Harvard University Press, 1999).

tickets, of the ratio of winning to losing tickets). Bayes's solution to this problem, and the work built upon it, starts, in contrast to frequentism, with an interest in probability as a subjective method of belief. This calculation of subjective belief was largely rejected by frequentists and the majority of mid-twentieth-century statisticians, but in recent years it has become increasingly persuasive and useful.

We will turn to more specifics shortly, but we should note that, if Ronald Fisher modeled a statistics for an agrarian society, and Jerzy Neyman and Egon Pearson modeled one for an industrial society, Bayes—through his modern interpreters—has provided a statistical theory for the information age. The use of Bayesian statistics has exploded in recent decades as statisticians strive to address the methodological shortcomings of frequentist approaches and take advantage of cheap computing power.

To fully appreciate the revolutionary implications of the rise of Bayesian approaches, it is best to start with frequentism's shortcomings. As the previous chapter outlined, problems with the approach have appeared in the sciences writ large, and statisticians, in their own ways, have also taken aim at the theoretical underpinnings of frequentism. For example, in 1976, statisticians Dennis Lindley and Lawrence Phillips demonstrated one of the problems with frequentism's claims to objectivity with a simple imagined experiment designed to determine if a coin is biased toward heads.[2] In their thought experiment, a researcher flips a coin twelve times, resulting in hhhthhhhthht, or three tails and nine heads. Assuming that the coin has an equal probability of landing heads or tails (the null hypothesis in this experiment), we can calculate the odds of seeing three or fewer tails in a run of

---

2 Dennis V. Lindley and Lawrence D. Phillips, "Inference for a Bernoulli Process (a Bayesian View)," *American Statistician* 30, No. 3 (1976): 112–119; David J.C. MacKay, *Information Theory, Inference, and Learning Algorithms* (Cambridge: Cambridge University Press, 2003); Adam P. Kubiak, "A Frequentist Solution to Lindley and Phillips' Stopping Rule Problem in Ecological Realm," *Zagadnienia Naukoznawstwa* 50, No. 200 (2014): 135–145.

twelve flips, which works out to 7 percent.[3] With the standard p-value cutoff of 5 percent, we would be led to conclude that there is not enough evidence to reject the possibility that what is observed is due to chance; hence it is still possible that the coin is fair.

But if the experiment were designed differently, using the exact same coin and results, we could contrive a different probability. As Lindley and Philips suggest, this time the researcher could decide to flip the coin until three tails appeared—so that the variable of interest becomes the total number of flips instead of the number of tails. With this new experimental design, they find the probability of twelve or more total flips becomes just over 3 percent. This difference of probability arises due to the fact that our researcher is now calculating the probability of two or fewer tails in the first eleven flips (since the twelfth flip can either be a tail—and analysis would stop there because they would observe the stopping condition of three tails—or a head—in which case the total number would be more than twelve).[4]

The statistical result of 3 percent, then, appears significant at the 5 percent level, and our researcher can conclude that, unlike the first experimental design, there *is* evidence against the coin being fair. While, on the one hand, it is reasonable that the design of an experiment would affect the results, on the other, this example suggests that the researcher's state of mind in carrying out an experiment can fundamentally modify the conclusions drawn from the same set of data. Thus, this supposed objective theory of probability rests squarely on the personal and subjective understanding of the experimenter.

For this reason, reference classes—that is what set of events are considered part of the experiment—are pivotal for

---

[3] There are 299 sequences that contain 3 or fewer tails out of 2^{12} possibilities, which ends up being 7 percent.

[4] The probability then becomes 134 out of 4,096 possible combinations, or 3 percent.

frequentism. While probability for frequentism is deemed objective—in that it measures only the number of occurrences of an event from a long run of trials—the construction of this long run requires a subjectively assembled grouping. As is often the case in philosophy and science, claims of objectivity find that they must, in the final analysis, rest on some subjective ground.[5] For probability to become objective, it requires an imaginary, subjective and thus incredibly human subsidy. The decision of how to compile the reference class of events from which probability is computed can significantly alter the results of a statistical experiment or analysis.

We witness here, again, the torsion between the subjective and the objective that we have been tracing. The more any perspective tries to get to the objective, real heart of the matter, the more quickly one slips back into the subjective. And, just as with capitalism, the cost of being a hard-nosed realist is that one believes in the most imaginary of inventions, such as a near-infinite series of coin flips or the guaranteed value of money. The more one tries to imagine a world beyond subjectively produced, human abstractions, the more central these abstractions become.

The Inverse Problem and the Necessity of Subjectivism

We have already discussed a number of problems with frequentism, but it is helpful to lay out its two main challenges schematically. First, as discussed in the previous chapter, in their strict interpretation frequentist approaches do not permit the description of single events probabilistically. Leonard Savage criticizes these approaches, writing that "objectivistic views typically attach

---

5 For example, in regard to physics, see Karen Barad, *Meeting the Universe Halfway: Quantum Physics and the Entanglement of Matter and Meaning* (Durham, NC: Duke University Press, 2007).

probability only to very special events. Thus, on no ordinary objectivisitic view would it be meaningful, let alone true, to say that on the basis of the available evidence it is very improbable, though not impossible, that France will become a monarchy within the next decade."[6] This limitation in applicable events is due to the fact that frequentist probability requires an imaginary long run of events whose frequency of a certain outcome provides the measure of probability. Thus, probability requires this long run, as opposed to a single event.

Second, and perhaps the most damaging critique of frequentist approaches, is that frequentism answers the wrong question. Probability is relatively easy to calculate if one knows everything about a system that is being modeled. If a statistician knows that a coin flip is fair, it is trivial to calculate the odds of getting two heads in a row. The probability is 0.25: the result of multiplying 0.5 (the chance of heads on the first flip) by 0.5 (the chance of heads on the second flip). But it is another matter altogether to go the opposite direction: to calculate the probabilities in the underlying system from observed data.

Frequentists throw up their hands when faced with this challenge. Instead, they ask a much more circumspect question: If one assumes a coin is fair, what is the probability that one would observe a given set of results? This is precisely the way that Fisher proposes hypotheses should be tested. While the answer to this question may offer some insight into the true frequency of the coin, asking what we would observe were the coin to be fair is neither philosophically nor mathematically the same question as asking what the probability is that the coin *is* fair. The frequentist pretense is based on an imagined world where the null hypothesis is true. Ultimately, as a result of the first problem, frequentism denies that it is possible to assign a probability to an actual hypothesis (since its truth is effectively a single event), offering instead the

---

6 Leonard J. Savage, *The Foundations of Statistics*, 2nd ed. (New York: Dover, 1972), 61–62.

probability that the evidence we witness would occur in a world where the null hypothesis is true.

Savage sums up the stakes of these developments well:

> In some respects Bayesian statistics is a reversion to the statistical spirit of the eighteenth and nineteenth centuries; in others, no less essential, it is an outgrowth of that modern movement here called classical. The latter, in coping with the consequences of its view about the foundations of probability which made useless, if not meaningless, the probability that a hypothesis is true, sought and found techniques for statistical inference which did not attach probabilities to hypotheses. These intended channels of escape have now, Bayesians believe, led to reinstatement of the probabilities of hypotheses and a return of statistical inference to its original line of development.[7]

Bayesian approaches thus offer to solve the inverse problem: rather than assume the truth of the null hypothesis, Bayesian statistics are able to offer a probability for the truth of the hypothesis itself, continually updating this probability as new evidence is gathered.

The Value of Bayes

Where the frequentism of Fisher, Neyman and Pearson interprets probability as representing a near-infinite run of a physical system—such as a series of coin flips—Bayesian approaches understand probability as a measure of subjective belief. Correspondingly, most frequentist approaches tend to think in terms of discrete

---

[7] Ward Edwards, Harold Lindman and Leonard J. Savage, "Bayesian Statistical Inference for Psychological Research," *Psychological Review* 70, No. 3 (1963): 193–242. See also, for example, Paul Pharoah, "How Not to Interpret a P Value?" *Journal of the National Cancer Institute* 99, No. 4 (2007): 332–333.

experiments, while Bayesians are more comfortable evaluating any and all data that is available. The Bayesian approach is novel because it lays out a method for making predictions that is well suited to the addition of new evidence.

With Bayes's formula, the specifics of which will be outlined shortly, new data can continually be added to the output of prior calculations. Due to this flexibility—an abandonment of rigid tests of hypothesis in favor of a constantly updated measure of belief—Bayesian statistics have seen a resurgence in the last few decades across both academic and corporate realms.[8] Bayesian approaches have been a boon to data-driven tech companies, as well as older manufacturing industries, as these established behemoths pursue new strategies for increasing profit in the milieu of contemporary informational capitalism.[9]

For a 2000 article in *Wired*, Michael Lynch, the founder of the data analytics company Autonomy—which has since been bought for $10 billion—declared that "Bayes gave us a key to a secret garden. A lot of people have opened up the gate, looked at the first row of roses, said, 'That's nice,' and shut the gate. They don't realize there's a whole new country stretching out behind those roses. With the new, superpowerful computers, we can explore that country."[10] In the intervening years, many companies have opened this gate, and they have profited handsomely by exploiting the resources of this new country, predicting everything from shopping habits to the outcomes of elections.

Early work in probability and statistics, especially Laplace's nineteenth-century development of what is now known as Bayes's theorem, made significant advances on this "inverse problem," calculating the probability of hypotheses themselves. However, this project was largely abandoned in the early twentieth century,

---

8 It is often forgotten how prevalent early versions of Bayesian approaches were prior to Fisher. His book *Statistical Methods for Research Workers* even begins with a refutation of "inverse probability."

9 Nick Srnicek, *Platform Capitalism* (New York: Polity, 2017).

10 Steve Silberman, "The Quest for Meaning," *Wired*, February 1, 2000.

due, in part, to Fisher and his followers' unease with the subjective nature of Bayesian statistics, as well as the tedium of the required calculations.

Bayesian approaches—unlike frequentist ones—require the use of what is called a prior probability distribution, which describes what one thinks to be the odds of various outcomes.[11] The use of this subjective prior probability bothered many of the early "scientific" frequentists, who hoped they could escape the need for subjective input in the process of statistical inference. Fisher was perhaps the most adamant in this regard, stating, "The theory of inverse probability is founded upon an error, and must be wholly rejected."[12]

As a whole, frequentists did not necessarily refuse the possibility of making statements about hypotheses. But, as Fisher argues, even though it may be possible to evaluate inferences (e.g., the truth of a hypothesis from data), probabilities cannot be directly assigned to a hypothesis:

> The rejection of the theory of inverse probability was for a time wrongly taken to imply that we cannot draw, from knowledge of a sample, inferences respecting the corresponding population. Such a view would entirely deny validity to all experimental science . . . the mathematical quantity which usually appears to be appropriate for measuring out order of preference among different possible populations does not in fact obey the laws of probability.[13]

---

11 Stephen E. Fienberg, "When Did Bayesian Inference Become 'Bayesian'?," *Bayesian Analysis* 1, No. 1 (2006): 1–40; Sharon B. McGrayne, *The Theory That Would Not Die: How Bayes' Rule Cracked the Enigma Code, Hunted Down Russian Submarines, and Emerged Triumphant from Two Centuries of Controversy* (New Haven, CT: Yale University Press, 2012).

12 Ronald A. Fisher, *Statistical Methods for Research Workers* (Edinburgh, UK: Oliver & Boyd, 1934), 9.

13 Ibid., 10.

Despite claims that it is able to make sweeping evaluations of hypotheses, frequentist hypothesis testing, when appropriately applied and interpreted, can only make very limited statements about them.

The combination of cheap computing power and practical problems with frequentism has led to a renewed interest in Bayesian approaches and growing comfort with subjective theories of probability. Because of its ability to employ prior belief, Bayes's theorem excels in conditions where we might want to update beliefs as data is gathered in real time—an especially exciting prospect for networked digital capitalism.

Bayes's Theorem

The heart of Bayesian analysis is known as Bayes's theorem, or Bayes's formula. Bayes's theorem states that the probability of an event (B), given that another event (A) has occurred, is the probability of A occurring—given that B has occurred—times the probability of B all divided by the probability of A occurring:

$$\mathcal{P}(B|A) = \frac{\mathcal{P}(A|B)\mathcal{P}(B)}{\mathcal{P}(A)}$$

In the context of frequentism, it is possible to see how Bayes's theorem allows us to treat "inverse probability." If we think of B as the truth of a hypothesis or theory and A as the evidence, this formula allows us to mathematically convert the probability of the evidence, given the truth of the hypothesis, to the probability of the hypothesis given the evidence—an inversion that frequentism explicitly rejects.

To understand how this works with a more practical example, imagine a medical test that can detect a disease 95 percent of the time, with a 1 percent false-positive rate. We may initially assume that if a subject tests positive, there is only a 5 percent chance they do not have the disease. Bayes's theorem can be used to calculate the probability based on the prevalence of the disease among the population. Bayesian inference starts with a subjective assessment (let us say the disease is rare; so, we assume only 2 percent of the population has it) and, as more evidence is obtained (the results of testing), the probability becomes increasingly accurate. With our initial assumptions, it turns out that a positive test corresponds with only a two-in-three chance of actually having the disease, despite the 95 percent rate of detection and 1 percent false-positive rate. This result may seem counterintuitive, but it is in fact a direct consequence of our assumption about the disease's prevalence: there are many more people without the disease (who could produce a false positive), than there are with it (who could produce a true positive).[14]

This example is a single test and is based on our belief in the prevalence of the disease in the population, since the true percentage is unknown. However, we could imagine testing a large population, then using these same calculations to update our presumed prevalence. This is the power of Bayesian approaches: in a world of cheap computation and massive amounts of digital data, we can continually update our beliefs about the world as new data arrives.

---

14 Using Bayes's theorem, the math is as follows. In this case, B is the probability that someone has the disease; A is the probability that we have a positive test. The only thing that is not immediately known is the probability of A, but that is simply the probability of a true positive (times the percentage of those with the disease) plus the probability of a false positive (times the percentage without the disease):

$P(A)$ = probability of a positive test for someone with the disease * percentage of people with the disease + probability of false positive * percentage of people without the disease

$P(A) = 0.95 * 0.02 + 0.01 * 0.98$

$P(A) = 0.0288$

The rest of the information we already know. The probability of A given B is the accuracy (0.95) and the probability of B (.02):

$P(B|A) = 0.95 * 0.02 / 0.0288$

$P(B|A) = 0.66$

Naive Bayes

With this in mind, let us turn to an example to demonstrate the modern power and relevance of Bayesian approaches: the field of information retrieval, where many celebrated machine learning algorithms have been nurtured. In particular, let us look at what machine learning researchers call the "naive Bayes classifier," an algorithm for classifying documents (or really anything about which data is known) into categories. It provides an exceptional example of how beneficial Bayesian approaches can be to modern data analysis.

This method is called "naive" because one assumes that the appearance of every word in a document is independent from every other (e.g., seeing the word "cat" does not affect the probability of seeing the word "dog" in the same document). This may seem counterintuitive based on what we colloquially know about language, as "cat" and "dog" are clearly more likely to appear together than, say, "cat" and "monad." But, as has been learned over the last few decades, many machine learning algorithms function exceptionally well if they are programmed naively, such that the algorithm can find structures and patterns on their own, rather than requiring guesses from programmers, who would then spend substantial amounts of time and effort trying to intuit which relationships are the most important.

In the case of naive Bayes, a programmer takes a set of "training" documents that have already been classified. For this example, let us say they are interested in whether a given document is about animals or not. This training data already has an implicit classification taxonomy for all of the documents in the set: either they are, or are not, about animals.

Then, in the simplest terms, a naive Bayes classification algorithm goes through this training dataset and calculates how likely each term is imagining that a document is in each of our classes in turn (we may discover that in an article about animals, the word "dog" has a 20 percent chance of occurring, while in a nonanimal article it has a 2 percent chance of occurring).[15] In essence, it computes the probability of a word for each category. Then, when the algorithm has fully passed through the training data, it can classify a new document (one not included in the training data) by calculating the inverse probability based on all of the words in the document. In short, the algorithm multiplies the probability of each word, for each category, by the overall probability of each category in our training dataset. The calculated result with the highest probability is the most likely, in this case "about animals" or "not about animals."[16]

With this example we can see two significant benefits of Bayesian approaches to our current informational capitalism. First, as noted earlier, the Bayesian approach allows for the integration of new evidence. After the classification algorithm categorizes a document or data element, that new categorization data can then be added to the probabilities for future documents (or if a human corrects the categorization, that can also be added). Moreover, while the above example dealt with the categorization of discrete and static documents, the same process can be used to categorize anything, updating as new data are added (e.g., predicted gender for a person's social media account can be recomputed after every post). With Bayes's theorem, we have an explicit, and hence automatable way, to continually add new data to our model.

Variations of the naive Bayes classifier have existed since the 1960s. While the algorithm for classification is relatively straightforward, it

---

15 Often a number of preprocessing steps and/or normalization are performed to make the analysis more accurate, e.g., stemming, term frequency-inverse document frequency and Laplace smoothing.

16 David D. Lewis, "Naive (Bayes) at Forty: The Independence Assumption in Information Retrieval," in *Machine Learning: ECML 1998*, Claire Nédellec and Céline Rouveirol, eds., *Lecture Notes in Artificial Intelligence*, Vol. 1398 (Berlin: Springer, 1998); Harry Zhang, "The Optimality of Naive Bayes," in *Proceedings of the Seventeenth International Florida Artificial Intelligence Research Society Conference*, Valerie Barr and Zdravko Markov, eds. (Menlo Park, CA: AAAI Press, 2004), 562–567.

is also incredibly powerful. Peter Norvig, Google's research director, has stated that "there must have been dozens of times when a project started with naive Bayes, just because it was easy to do and we expected to replace it with something more sophisticated later, but in the end the vast amount of data meant that a more complex technique was not needed."[17]

Second, we can begin to see how Bayesian analysis constructs the bridge between statistical hypothesis testing and the advent of machine learning. Using a naive Bayes classifier, a statistician is able to take data (such as word counts) and calculate the probability of each hypothesis (i.e., "animal" or "not animal"), which then allows them to decide the most likely one. Yet, from another perspective, that statistician is simply classifying documents and calculating the most likely category. And if they use a different training set of data the next time they run the algorithm, they may even change their mind about a document. By doing away with the necessity for discrete experiments, Bayesian analysis allows hypotheses to multiply nearly infinitely: each document, each visitor to a website, each loan application appears as a hypothesis whose probability can be calculated and from whose outcome more can be learned.

By assigning a probability to hypotheses, Bayesian approaches are able to automatically and computationally weigh different hypotheses. At first glance, the subjectivism of Bayesian approaches may appear to demand human involvement. But in reality, this computational subjectivity actually empowers computers because it provides a concrete mathematics that they can follow. With this formulation, an algorithm can exploit its subjective (i.e., local) position to calculate probabilities and does not need to rely on a human who can understand a transcendental totality (i.e., the assigned reference class or experimental design in frequentist analysis) that allows and requires them to

---

[17] Peter Norvig, quoted in McGrayne, *The Theory That Would Not Die*, 244.

arbitrate objectivity, as is the case with frequentism (e.g., the need to design an experiment, select a reference class and interpret the results). This capacity allows computers, without human input, to choose the most likely cause for a set of observations in a way that strict adherents of frequentism were never comfortable with.

This is one of the critical changes ushered in by the Bayesian revolution: the ability to assign a probability to a single event, including a hypothesis, means the process can be automated in a way that is not allowed under a strict frequentist interpretation of probability, precisely because in the latter, a hypothesis cannot be assigned a meaningful probability. Once a hypothesis *can* carry a probability of being true, each possible classification can be assigned a probability and the most probable selected. Thus, even if Bayesian probability is subjective, the subject that evaluates it can just as easily (or often more easily) be a computer rather than a human. And from this subjective perspective, computers are able to automatically compute the most valuable option. In this way, as we shall see in more depth, this Bayesian approach takes up the economic commitment of Neyman and Pearson and commits itself to a general management of knowledge and society through probabilistic reason. In form, this Bayesian method finds extraordinary harmony with the profit-seeking dictates of automated information capitalism.

While humans are still intimately necessary when it comes to defining a problem and determining what a valuable result is, Bayesian approaches make a revolutionary advance toward the objectification of parts of this process, and thus toward its computability. Ultimately, the Bayesian revolution is a revolution in capitalist production that may one day come to be seen as important as the Taylorist revolution, automating and accelerating knowledge production just as Taylorism did industrial production. Bayesian statistics has fundamentally altered the production of knowledge, allowing the mechanization of a process that turns data into generalizable facts. Even beyond strict Bayesian approaches, the larger change in method it has precipitated allows probabilistic insights to be brought directly to market, where they automate decision making based on data and market demands. And, as we shall shortly see, these methods integrate market valuations directly into the metaphysics of knowledge production.

Chapter 7
Bayesian Metaphysics and the Foundation of Knowledge

Bayesian approaches flip frequentism's metaphysical perspective on its head. Instead of starting with an objective theory of probability only to end up having to rely on imaginary, subjectively created reference classes, Bayesian approaches start with a subjective belief and slowly, but procedurally, move toward objectivity. Leonard Savage explains the process through the example of a coin:

> Although your initial opinion about future behavior of a coin may differ radically from your neighbor's, your opinion and his will ordinarily be so transformed by application of Bayes' theorem to the results of a long sequence of experimental flips as to become nearly indistinguishable. This approximate merging of initially divergent opinions is, for Bayesians, how empirical research becomes "objective."[1]

In a certain sense, the Bayesian approach takes the process of objectification more seriously than frequentism; it simultaneously

---

1 Ward Edwards, Harold Lindman and Leonard J. Savage, "Bayesian Statistical Inference for Psychological Research," *Psychological Review* 70, No. 3 (1963): 193–242.

provides a means of making individual and socially situated knowledge objective *and* provides a logic that attempts to show the subject how they should act under the reign of its metaphysics.

In this metaphysics, we can hear the mantra of objectification: do and think what you want, but in the end we must all be realists and sell our labor and compute our probabilities if we want to eat. Here the Bayesian approach produces its "real abstraction." While Bayesian analysis turns to a subjective account of statistics, as we will see, it ultimately grounds the rationality of this subjectivity in the exchange of contracts. Despite the hyper-appropriateness of this means of knowledge production to capitalist exchange, by revealing the social nature of knowledge production, Bayesian analysis points beyond the limitations of capitalism and ultimately demonstrates the very antipathy of capitalism to knowledge production.

Percentage Belief

Instead of imagining, and then defining, some group that becomes the reference class that is being sampled from, Bayesian approaches imagine the researcher or the computer as an agent that continually gains more knowledge of the world. In some cases, these probabilities end up being similar to frequentist analysis and can even behave similarly (e.g., if we have a 50 percent belief that a coin flip will return heads, we expect that over a long run of actual flips, 50 percent will tend to be heads).[2] But the interpretation, use and metaphysical grounding of these probabilities fundamentally differ.

---

2 Another major difference between Bayesian and frequentist statistics is the former's general preference for posterior distributions rather than point estimates of statistics (e.g., mean). Mathematically and computationally, this shift is of the utmost importance. It is worth noting that the use of distributions aids in the calculation of uncertainty and ability to represent possible but unlikely events, further cementing Bayesian approaches' relevance to our present moment. Bradley P. Carlin and Thomas A. Lewis, *Bayes and Empirical Bayes Methods for Data Analysis*, 2nd ed. (Boca Raton, FL: Chapman & Hall / CRC, 2000), 12.

While this use of subjective probabilities may seem like an antiscientific step backward—an abandonment of objectivity and reversion to our initial uninformed, or minimally informed beliefs—Bayesian analysis has gained a strong foothold in the scientific world. This is partly due to the decreasing price of computing power, but also because Bayesian analysis has allowed novel approaches for thinking through key issues of scientific method. For example, Bayes presents a potential means for understanding the replication crisis and problems in scientific inference, since the use of a prior probability now allows observers to demand extraordinary evidence for unlikely claims: if one were to think, say, that the existence of extrasensory perception is highly improbable, massive amounts of evidence would be required to overcome this prior belief against its existence. Furthermore, any known biases—such as the increased probability that researchers will publish surprising results or selectively report data—can be factored into the analysis of the larger scientific consensus on an issue. In short, with Bayesian analyses, one can mathematically model not just a single experiment, but the entire state of a given field of knowledge.[3]

Even more important, the Bayesian approach fundamentally changes the stakes of statistical analyses. As a result of its subjectivism, the result of a statistical test no longer claims to tells us the "truth" of the matter. Rather, each result can be taken as evidence so that each reader can calculate their own expected probability. For instance, one researcher may believe the existence of extrasensory perception is highly unlikely, while another may think its existence is equally as likely as its nonexistence, and each would

---

3 This array of larger states of various fields of knowledge are called meta-analyses. These meta-analyses have a relatively long history, going back to at least Karl Pearson's frequentist work, which held formidable effect on the practice, at the very beginning of the twentieth century. This 1904 paper is generally taken as the first modern meta-analysis: Karl Pearson, "Report on Certain Enteric Fever Inoculation Statistics," *British Medical Journal* 2, No. 2288 (1904): 1243–1246. While frequentists pioneered much of this work, Bayesian approaches offer the possibility of pushing them even further to better reflect the state of fields and belief in their discoveries.

calculate a different probability after reading an article providing some evidence of its existence. This focus on belief short-circuits some of the problems with frequentist statistics: scientific studies no longer need to pretend that they provide us with a transcendent truth. Instead, these studies offer data to readers, which they can then use to update their own beliefs.

And through this constant overhaul of knowledge, Bayesian statistics provides methods and an ideology, both of which are well suited to informational capitalism.[4] In so doing, it provides a mathematical framework in which evidence can be transformed into belief, which can then be transformed directly into automated decision making (e.g., about what ad to show a visitor to a website, determined by treating each possible ad as a hypothesis and selecting the one with the highest probability of getting clicked). Second, this shift to individual belief creates markets from which the combination of evidence (data), and models (a company's simulation of the way the world or market likely works) can provide a competitive advantage in the marketplace against others with lower-quality data or assumptions.

We can see the potential for this marketization explicitly in the case of the "animal" and "not animal" naive Bayes classifier from the previous chapter. If enough high-quality training data is used, the computed model becomes incredibly valuable because it can

---

4 In an often-cited footnote found in the fifteenth chapter of Volume 1 of *Capital*, titled "Machinery and Large-Scale Industry," Marx writes: "Technology reveals the active relation of man to nature, the direct process of the production of his life, and thereby it also lays bare the processes of the production of his life, and thereby it also lays bare the process of the production of social relations of his life, and of the mental conceptions that flow from those relations." *Capital*, Vol. 1 (London: Penguin Books, 1990), 493n4. Given Marx's emphasis on the historical specificity of social relations to a given mode of production, this statement clearly indicates the extent to which the technologies we use today are immanent to and revelatory of the social relations that make possible the valorization of capital in our current historical moment. Gilles Deleuze makes a similar argument when he writes, "Types of machines are easily matched with each type of society—not that machines are determining, but because they express the social forms capable of generating them and using them." "Postscript on the Societies of Control," *October* 59 (1992): 6.

allow a company, scientist or government to make predictions, and even the most miniscule of increases in accuracy directly supplies a competitive advantage. In this way, the Bayesian revolution provides a key set of methods for informational capital, allowing the computation of knowledge from data.

The Market as Metaphysical Ground

Like probability in general, the origins of Bayesian analysis admit to a theistic understanding of probability that, despite being largely abandoned, still speaks to its metaphysical power. Following Bayes's death, his friend Richard Price discovered an intriguing manuscript among Bayes's papers. After two years of significant editing, Price published the manuscript as *An Essay towards Solving a Problem in the Doctrine of Chances*. In his introduction to the text, Price makes a profound comment about the theological importance of Bayes's discovery:

> The purpose I mean is, to shew what reason we have for believing that there are in the constitution of things fixt laws according to which things happen, and that, therefore, the frame of the world must be the effect of the wisdom and power of an intelligent cause; and thus to confirm the argument taken from final causes for the existence of the Deity. It will be easy to see that the converse problem solved in this essay is more directly applicable to this purpose; for it shews us, with distinctness and precision, in every case of any particular order or recurrency of events, what reason there is to think that such recurrency or order is derived from stable causes or regulations in nature, and not from any irregularities of chance.[5]

---

5 Richard Price, introduction to Thomas Bayes, "LII: An Essay Towards Solving a Problem in the Doctrine of Chances. By the Late Rev. Mr. Bayes, F.R.S. Communicated by Mr. Price, in a letter to John Canton, A.M.F.R.S," *Philosophical Transactions* 53 (1763): 370–418.

For Price, the very predictability of the world—even if it is probabilistic—proves the existence of "regular laws" and hence divine design.

Thus, in the face of this problem, Price suggests that it is not the existence of the divine that proves the laws of the universe, but rather the inverse: the existence of these regularities proves the existence of the divine. In this sense, Price's view is similar to that of John Arbuthnot, who performed one of the first known statistical tests and believed that if what is observed is not due to chance alone, then God's existence and intervention is verified.[6]

Price's argument suggests a valuable direction for our present concerns: calculations of belief bear within them a fundamentally theological commitment. It should be remembered that Bayes was a man of the cloth, and one of the two essays he published during his life attempted to prove that God wants us to be happy.[7] Here we see the vital, divine connection between a statistical, subjective measure of belief and proof that a transcendent force watches over the world—or a system—ensuring its proper functioning.[8] While this subjective force found its justification in a transcendental, otherworldly anchor, today it appears to function even without that locus.

Indeed, one of the barriers many statisticians find to accepting Bayesian metaphysics is that it remains difficult to argue for a specific, universal set of rules for determining subjective belief. For example, if probability is subjective, how can one effectively determine that a researcher's belief is more or less true than another's? For subjective probability to be efficacious in its predictive capacity, there must be some objective means of deriving, describing and

---

6 John Arbuthnot, "An Argument for Divine Providence, Taken from the Constant Regularity Observ'd in the Births of Both Sexes," *Philosophical Transactions* 27 (1710): 186–190.

7 Thomas Bayes, *Divine Benevolence; or An Attempt to Prove That the Principal End of the Divine Providence and Government Is the Happiness of His Creatures* (London: John Noon, 1731).

8 Jean-Pierre Dupuy, *The Mark of the Sacred* (Redwood City, CA: Stanford University Press, 2013).

computing this subjective belief—and thus for binding that belief to the laws of probability and statistical induction.

This problem of where subjective probability attaches itself to the laws of the universe becomes all the starker when we remove the epistemic, transcendental anchor of a god from the situation. According to Bayes and Price, the discovery of these new mathematical laws allows one to witness the regularity and mathematical design by which God laid out the universe. For them, it likely does not matter if others do not share their belief or think by their theological rules. It is sufficient that they know and God knows. We see here again the structure of objectified belief, with God as the force of objectification: this is how things work, with or without one's individual belief. But for those mathematicians who do not believe in a god, subjective probabilities must find their ground elsewhere; and it is ultimately the market that comes to provide this "elsewhere," and with it the full capitalist force of contemporary objectification.

Dutch Book Argument

While multiple attempts to construct justifications for Bayesian probability have been offered, one of the most classic and enduring explanations is the "Dutch book argument." This argument was introduced in a 1937 article by the Italian statistician and actuary Bruno de Finetti.[9] A Dutch book is a gambling situation in which, by taking a given bet (or set of bets), a gambler is guaranteed to lose. De Finetti demonstrated how the mathematics for dealing with probabilities can be derived from a subjective position in which one's only motivation is to avoid having a Dutch book made

---

9 Bruno de Finetti, "Foresight: Its Logical Laws, Its Subjective Sources," in *Studies in Subjective Probability*, Henry E. Kyburg and Howard E. Smokler, eds. (New York: Wiley, 1964), 93–158. Frank Ramsey hinted at this solution in his "Truth and Probability," in *Readings in Formal Epistemology* (New York: Springer, 2016 [1931]), 21–45. However, the Dutch book argument was fully developed by de Finetti.

against them. Here, probabilities are converted into the price (or odds) an agent would take for buying a contract. For example, if that individual believed a coin was fair, they would be willing to pay even odds for a bet on heads (i.e., a one-dollar bet would pay out an additional dollar for a correct guess). But if they thought that coin was biased, and landed heads twice as often, they would require a corresponding payout of two to one to bet on tails.

From the aim of avoiding a Dutch book, it is possible to derive the basic mathematical laws of probability. For example, the sum of the probability that an event (e.g., it rains tomorrow) happens and the probability that it does not happen should not be greater than one. But if we accept odds of one to one that it happens, and two to one that it does not happen, then someone else could bet three dollars that it will rain and two dollars that it will not. If it rains tomorrow, we will win three dollars on the first bet but lose four dollars on the second—a net loss of one dollar. Conversely, if it does not rain, we will lose three dollars on the first bet and gain two dollars on the second—again, a net loss of one dollar. Whatever happens tomorrow, we lose. A Dutch book has been made against us. If we abstract from this example, it becomes possible to prove that the probability of an event and its opposite must sum to one.[10] Likewise, the other fundamental rules of probability can be derived.

Most important for our current purpose, the Dutch book argument offers an economic grounding for the mathematical laws of probability, even in the subjective realm of Bayesian probability. Tellingly, while Ronald Fisher, Jerzy Neyman and Egon Pearson were busy turning the academic world toward frequentism in the early and mid twentieth century, actuaries—like Arthur Bailey, who kept Bayesian approaches alive during the heyday of frequentism—had to be "realists" about costs of risks in calculating

---

10 See Dennis Lindley, *Understanding Uncertainty* (Hoboken, NJ: John Wiley & Sons, 2013), esp. Section 5.7.

insurance rates and thus refused to abandon Bayesian methods.[11] Such "realism" confirms the capitalist suspicion that exchange and prices can serve to translate subjective intuition into the objective truth of the market.

In Bayesian analysis, according to the Dutch book argument, calculation takes on the form of contract exchange and thus makes apparent the structure of objectification: subjective belief is tethered to the objective conditions of exchange.[12] We can believe

---

11 Sharon B. McGrayne, *The Theory That Would Not Die: How Bayes' Rule Cracked the Enigma Code, Hunted Down Russian Submarines, and Emerged Triumphant from Two Centuries of Controversy* (New Haven, CT: Yale University Press, 2012).

12 A funny thing happens: the quicker we run "to the things themselves," the faster we end up dealing with subjective theories of knowledge, or, at the very least, with theories that must first deal with the subject, as Edmund Husserl brilliantly demonstrated. In this light, what should interest us is not how things are subjective or how they are objective, but rather the techniques by which the subjective and objective are sutured together; how subjective knowledge is made to appear objective; in short: objectification.

We can see this explicitly in object-oriented ontology. Graham Harman, in his careful ontological analysis, ends up insisting that the outcome of philosophy is that we have to pay and pay. For Harman, all philosophical assumptions must be brought to market and the proper price paid for their intellectual assumptions. For instance, he says of Giordano Bruno's philosophy, "The technical term for this maneuver is 'highway robbery,' since Bruno is trying to preserve individual forms without paying for them." "On the Undermining of Objects: Grant, Bruno, and Radical Philosophy," in *The Speculative Turn: Continental Materialism and Realism*, Levi Bryant, Nick Srnicek and Graham Harman, eds. (Melbourne: re. press, 2011), 35. Elsewhere, Plato and Socrates must "pay the immense price of reducing actors as we know them to flickering shadows on a cave wall." *Prince of Networks: Bruno Latour and Metaphysics* (Melbourne: re. press, 2010), 95. While the use of this turn of phrase is not damning evidence on its own, it should give us pause that a philosophy based on the solidity of objects advances through the keeping of accounts on who has to pay for what. This is precisely what objectification in Marx's sense suggests: objects serve to account for the debts and credits of obfuscated social relations.

Bryant's text on object-oriented philosophy is particularly exemplary in this regard. The title, *Democracy of Objects* (London: Open Humanities Press, 2013), a phrase he draws from Bogost, is striking; especially in so much as he dismisses representation as a mere epistemological question, despite representation being the sine qua non of contemporary democracy. In the introduction to the text, Bryant talks at length about representation being a central question to both realist and anti-realist philosophies, which object-oriented ontology is able to sidestep by seeking to

whatever we want, but there is only one way to avoid being cheated. Bayesian analysis no longer needs a god to guarantee the ground of belief: the Dutch book now takes the place of that other book that surely guided Reverend Thomas Bayes.

Just like Marx's commodity, the Dutch book inscribes objectivity into the heart of belief, and in so doing provides an exemplary

"reject the epistemological realism of other realist philosophies, taking leave of the project of policing representations" (26–27), a step which he says allows philosophy to pluralize the gap between object and representation. While I am sympathetic to this desire to pluralize representation and to consider the real beyond representation, representation remains the central question of democracy today. Especially if we understand that representation in its very function is a process of exclusion, it seems that a democracy of objects (or of anything) must provide a singular theory of representation. Even if we admit a plurality of possible "representations," democracy must select one and police its functioning. One vote per object over the age of eighteen; polling stations are only open from 7 a.m. to 8 p.m.!

Bryant clearly has something very different in mind when he uses the term "democracy," than that meant by the political question of democracy and its future that confronts us today. For Bryant, this is "a democracy of strange strangers. Where there is no hegemon that stands above and outside withdrawal as a full actuality, there is only a flat plane composed of strange strangers" (269–270). While we can sympathize with these anti-hegemonic desires, it is odd to call it a democracy if there is no hegemonic function to represent the totality. And Bryant goes on to deny that there is any "world," as a unifying function for the totality. But what could be meant by democracy if there is no world function to, even if out of nothing, create a world, a politics or a demos? Even if the real is flat, the demos, the people, must be constructed and represented. We must conclude instead that what Bryant claims to mean by "democracy" is in reality an anarchy of objects, a Hobbesian war of all against all, in short, the idealized market of capitalism with no democratic oversight.

Much more can and has been said about the relationship between democracy and capitalism, but for the time being, we will have to satisfy ourselves with the observation that Bryant's democracy of objects is a democracy without politics or representation. It is a democracy truly of the market; it is an objectified democracy, where the only politics are the politics of individual competitive relations between objects at market. Bryant's attempts to separate the ontological from the political precludes any understanding of the ways in which the global system of computation or capital are reflected in the very ontological distinctions that are made and thus the analysis replicates the very structure of commodity relations. While Bryant's text is only one cautionary tale, we can see in it the very real risks of ignoring the fundamental mysteries at the heart of these relations and the relevance still today of Marx's writings on object-commodities.

case of the process of capitalist objectification, removing any locatable ground and distributing truth into the market. Here, we see a series of social relations—under the threat of having a book made against oneself—force the subject to act in accordance with those supposedly "objective laws." Indeed, for de Finetti, even objectivity in science becomes a condition not of the regularity of the material world but of the predictability of our thoughts. He writes, "I do not look for why *the fact* that I foresee will come about, but why *I do* foresee that the fact will come about. It is no longer the facts that need causes; it is our thought that finds it convenient to imagine causal relations to explain, connect and foresee the facts."[13] Everything now appears subjective, but this subjectivity is tied to a contract market, which in the end requires that all who arrive at market act objectively.

De Finetti explicitly argues that his is an anti-metaphysical position, completely naturalizing itself and denying its wholly metaphysical suppositions. It finds its truth not in the sensuous world, but, just like capitalism, in an imagined ideal world that is claimed to be more objective and more true than our material existence. Math becomes the truest of sciences because it forsakes the existence of the world: "Mathematics, logic, and geometry are now immune to the pseudo hypothesis (so to speak) of the existence of the world, the existence of an external reality, the existence

---

13 Bruno de Finetti, "Probabilism: A Critical Essay on the Theory of Probability and on the Value of Science," *Erkenntnis* 31 (1989): 170. While in his later life, de Finetti moved to the left, even being arrested briefly for publishing letters of conscientious objectors in a newspaper for the Italian Radical Party, in the 1930s when he was developing and outlining his theories of probability he was a supporter of fascism. He concludes a short essay on the nature of scientific work by proclaiming: "That impeccable rational mechanics of the perfect civilian regime of the peoples, conforming to the rights of man and various other immortal principles! October of '22! It seemed to me I could see them, these Immortal Principles, as filthy corpses in the dust. And with what conscious and ferocious voluptuousness I felt myself trampling them, marching to hymns of triumph, obscure but faithful Blackshirt" (219).

of a metaphysical reality."[14] Through objectification, everything is turned upside down. To live in this objective world means to attend to the socially situated and economically mediated thoughts of humans, whereas to attempt to understand facts in the material "real world" is to engage in pure metaphysical speculation. The turn to a nonmaterial mathematics allows for a universal that cannot be assailed by any given particular. In this way, the subjective grounds of this mathematical approach produces a metaphysical, objective world: one that is simultaneously transcendent and bound to the universal laws of probability governed by the looming fear of the Dutch book.

While Leonard Savage does not make such strong claims about the nonexistence of the material world, he goes further than de Finetti to align subjective probabilities with the logic of the market. In his famed book, *Foundations of Statistics*, Savage returns to Neyman's work on behavioral understandings of statistics. He states that "the problems of statistics were almost always thought of as problems of deciding what to say rather than what to do, though there had already been some interest in replacing the verbalist by the behavioralist outlook. The first emphasis of the behavioralistic outlook in statistics was apparently made by J. Neyman."[15] Savage goes further commending the behaviorist approach of Neyman and Pearson as the ground of Bayesian statistics, stating, "Personalistic statistics appear as a natural late development of the Neyman-Pearson ideas."[16] In praising Neyman's behavioralism, Savage places the subjective theory of probability fully onto an economic foundation. In a sense, he reduces all epistemology to economics. While de Finetti abandoned only causality, Savage abandons knowledge altogether in favor of exchange.

---

14 Ibid., 171.
15 Leonard J. Savage, *The Foundations of Statistics*, 2nd ed. (New York: Dover, 1972), 159.
16 Ibid., iv.

Savage published *Foundations of Statistics* in 1954, yet in many ways it still represents the state of statistics' foundational metaphysics. To offer one example: Graciela Chichilnisky, the mathematical economist who developed the carbon credit trading model underlying the Kyoto Protocol (an exemplary case of an attempt to reduce all possible solutions to market logic), cites Savage's foundations favorably in her work, which focuses on evaluating the risk of highly unlikely but costly events, such as natural disasters and market crashes.[17]

Moreover, the substantial statistical advancements made in recent decades have built on the foundations developed by Savage, de Finetti and others of their generation. Challenges to the Dutch book argument have arisen in the intervening years, especially concerning inconsistencies that arise from the order in which bets are made, along with the apparent but impossible requirement that a gambler have "logical omniscience."[18] Yet the majority of these challenges argue for mere modifications to the argument rather than its wholesale rejection, leaving its foundation intact.[19]

Bayesians oversaw a revolution of science, and now, as researchers learn to use massive stores of data and cheap computing power

---

17 Graciela Chichilnisky, "The Foundations of Statistics with Black Swans," *Mathematical Social Sciences* 59, No. 2 (2010): 184–192.

18 See, for example, Alan Hájek, "Scotching Dutch Books?," *Philosophical Perspectives* 19, No. 1 (2005): 139–151; Ian Hacking, "Slightly More Realistic Personal Probability," *Philosophy of Science* 34, No. 4 (1967): 311–325; and Frederic Schick, "Dutch Bookies and Money Pumps," *Journal of Philosophy* 83, No. 2 (1986): 112–119. Hájek states: "The Dutch Book argument, in the form that has become a philosophical staple, is simply invalid, and to the extent that we have bought it for so many years we have, as it were, been Dutch Booked ourselves; indeed, in endorsing it we have been guilty of a certain form of incoherence. However, like Route 66, the argument can be reconstructed, or better still, restored" (139).

19 For a general overview of these concerns, see John Earman, *Bayes or Bust? A Critical Examination of Bayesian Confirmation Theory* (Cambridge, MA: MIT Press, 1996), 38–50. Earman ultimately concludes that the general Dutch book arguments, along with some other related arguments, are in the end convincing that the probability calculus can adequately describe subjective beliefs.

even more effectively, a new epistemic world is being built, piece by piece and study by study, on these new revolutionary approaches. As Andrew Gelman puts it: "If you wanted to do foundational research in statistics in the mid-twentieth century, you had to be a bit of a mathematician, whether you wanted to or not ... if you want to do statistical research at the turn of the twenty-first century, you have to be a computer programmer."[20] We should add to Gelman that in the first case, you had to have been a bit of a metaphysician, as well; now, one can build and program on top of the metaphysical foundations laid by Savage and his contemporaries.

Ultimately, we witness in Savage and de Finetti's work, and the broader turn to Bayesian statistics, a familiar challenge: that of objectification's torsion between the objective and subjective. The more we attempt to get behind the mask of how things "really" work, the more quickly we end up back at the subjective experience of social relations. Thus, as contemporary informational capitalism comes to require more and more finely tuned sources of objective knowledge, researchers and statistical metaphysicians discover only the hieroglyphics that capitalism leaves behind.

By founding statistics on contract exchange, we discover a contemporary example of Alfred Sohn-Rethel's claim that abstract thinking develops on the grounds of exchange. For Sohn-Rethel, the exchangeability of goods, and their ultimate representation as money, creates the material conditions in which incommensurate objects can face each other and be equated.[21] In tethering knowledge to exchange, the subjective and the objective can never

---

20 Andrew Gelman, "Bayes, Jeffreys, Prior Distributions and the Philosophy of Statistics," *Statistical Science* 24, No. 2 (2009): 176–178. Gelman here also explicitly states his antipathy to Savage's work: "I confess to having found Savage to be nearly unreadable, a book too much of a product of its time in its enthusiasm for game theory as a solution to all problems ... When it comes to Cold War-era foundational work on Bayesian statistics, I much prefer the work of Lindley." But, his concerns are more stylistic than substantive, especially considering that Lindley also makes recourse to Dutch book arguments.

21 Alfred Sohn-Rethel, *Intellectual and Manual Labour: A Critique of Epistemology* (Atlantic Highlands, NJ: Humanities Press, 1978).

completely abandon each other, for exchange twists the subjective and objective together. They must, rather, find their respective ground in their opposite, with objectivity produced out of the subjective position and the subject bound objectively to laws of exchange. Moreover, the development of the Dutch book argument speaks to the victory of neoliberalism as a means to organize all life and subjectivity on the grounds of market exchange.[22] Thus, historically we go from Fisher's individualism, to the managed objectivism of Neyman and Pearson, and finally back to the individual—only now, this individual is one who must follow the laws of capitalist exchange.

At first glance, it may appear that this shift in the foundation of statistics, from Bayes's Bible to de Finetti's Dutch book, has created a relatively final and stable form that seamlessly melds statistics with the necessities and ideology of capitalism. In sweeping away any ground for knowledge outside the exchange relationship—whether it be God or the objectivity of a long-run frequency—this formulation declares that only a well-informed exchange of contracts can underwrite knowledge. In essence, it claims that the only way forward is either the acceptance of this victory in favor of capitalism, or conversely, resistance to this economism at all costs, through opposition and critique.

But, at second glance, we can begin to see the contradiction at the heart of this metaphysics: the very incentives of the exchange on which this knowledge is founded call forth dissimulation rather than knowledge.[23] Value is always relative; so, the underlying goal of the exchange of contracts is not to increase knowledge

---

22 See Wendy Brown, *Undoing the Demos: Neoliberalism's Stealth Revolution* (Cambridge, MA: MIT Press, 2015).

23 Gerd Gigerenzer and Julian N. Marewski suggest that the flood of false results and faked data are a result of the belief in a universal statistical method following the "inference revolution" in the 1950s: "Surrogate Science: The Idol of a Universal Method for Scientific Inference," *Journal of Management* 41, No. 2 (2015): 421–440. This is likely correct, but it should be added that this desire for a universal method that can implemented mechanically is at least in part the result of contemporary modes of knowledge production.

absolutely, but rather to know relatively more than the other party to the contract. Especially as the ability to extract knowledge from data advances, it becomes increasingly difficult to gain such an advantage, incentivizing research that aims to stupefy rather than to add to collective or even individual knowledge.[24] The political economy in which these transactions occur encourages and rewards the accumulation not just of capital, but of knowledge—and with it the ability to construct reality such that the other party in the exchange does not know.[25] When statistics and science find their ground in the accumulation of capital, the incentives of capital accumulation will always override the desire for knowing, favoring relative knowledge and with it dissimulation.

Moreover, despite capitalism's redefinition of science, the world has not completely done away with Fisher's insistence on the lone scientist who knows for herself. This fantasy of probabilistic knowledge that precedes exchange obfuscates the threat that capitalism poses to science, as it allows one to maintain that some objective measure of probability stands outside of political economy. This leads some to believe that even if economic incentives distort processes of knowledge production, true scientific knowledge—through better methods or changing incentives within universities—can be dug out from under the detritus capitalism throws on top of it.

We can see here again the complex interplay between the abstract and concrete, with their attendant forms of domination. While Bayesian analysis may lend itself to the maximization of profit, what is discovered there is social reality (including its

---

24 Bernard Stiegler addresses this stupefying tendency directly in *States of Shock: Stupidity and Knowledge in the 21st Century* (Hoboken, NJ: John Wiley & Sons, 2015).

25 Manuel DeLanda, "Markets and Antimarkets in the World Economy," in *Technoscience and Cyberculture*, Stanley Aronowitz, Barbara Martinsons and Michael Menser, eds. (New York: Routledge, 1996) 181–194. In this essay, DeLanda argues for a synthetic approach to an analysis of capitalism that can demonstrate the historical emergence of monopolies as the natural outgrowth of capitalism rather than markets.

racism, sexism, etc.) presented in objective—that is, abstract—form. But then, to make matters even worse, this is often interpreted through a Fisherian—that is, concrete—lens, providing further force to what was initially discovered merely as a temporary, contingent effect of the market. Thus, statistics and algorithms effectively launder nonobjective forms of violence and bias, giving them greater stability, only to be fed back into these systems as the initial conditions for the next round.[26]

The discovery of the Dutch book argument may appear to put statistics on a solid ground through capitalist exchange, but in the final analysis we discover that scientific production and capitalism have now split ways. For the dictates of value extraction threaten, always and everywhere, to outrun and overturn the necessities of knowledge production, both though external incentive structures and internal metaphysical structures. If science is to have a future for humanity, it must be in opposition to and outside of capitalism. Under the conditions of late capitalism, we cannot go back to the halcyon days of Fisher, when science could claim to be a royal road to individually held truth. Science and the production of knowledge now require and call for a foundation that turns against capitalism.

This, then, is the ultimate task of a revolutionary mathematics today: to work toward the future of mathematics and science, redefining their underlying metaphysics, with a full understanding of the political and economic stakes that both determine and are determined by the possibility of this future. On a metaphysical level, the Bayesian revolution has overturned the individualist and chauvinistic Fisherian paradigm. The revolutionary mathematician, even if she is not a mathematician or a scientist in the colloquial sense, seeks to create new truths and new computations in the wake of this tumult. The future of science is a collective

---

26 See Wendy Hui Kyong Chun, "Queerying Homophily," in Clemens Apprich, Wendy Hui Kyong Chun, Florian Cramer and Hito Steyerl, *Pattern Discrimination* (Lüneburg: Meson Press, 2018), 59–97.

endeavor, and as such, our collective and social existence provides the very ground and possibility of the sciences.

By no means should this be taken as opposition to statistics, calculation or prediction. It is not necessary to find some surplus of language or some form of non-exchangeability in order to resist capitalism and computation.[27] On the contrary: computation and exchange, at their heart, work on the level of non-exchangeability. Statistics is nothing short of magic, performing metaphysical work that sutures our subjective and probabilistic knowledge to the material world. It mediates between the particular (data) and the universal (hypothesis), making the uncomputable computable. But in doing so, it functions within and through political economy and exchange.

While the knowledge that algorithms and statistical methodologies produce are often presented with a Fisherian veneer that claims recourse to some transcendent and individual truth, their actual metaphysical support is only tethered to this world by economic advantage and risk. They reflect not the world as it is, but rather the world *as it is profitable*. Accordingly, a revolutionary mathematics must aim not to show the world "as it is," but rather to recognize the necessity and importance of this objectification. Fisher's realization that Neyman and Pearson's behavioral interpretation of statistics required a science managed by a five-year plan or plant manager was incredibly prescient. Indeed, the production of knowledge through probabilistic systems, whether statistics or algorithms, now requires some form of social investment in this production, in opposition to the dictates of capital accumulation. The only question that remains is whether statistics is to be managed by plant managers—at pharmaceutical companies, tech companies, university research offices, and so on—or by some form of collective set against and beyond capitalism.

---

27 Franco "Bifo" Berardi, *The Uprising: On Poetry and Finance* (Los Angeles: Semiotext(e), 2012); Alexander Galloway, "The Poverty of Philosophy: Realism and Post-Fordism," *Critical Inquiry* 39, No. 2 (2013): 347–366.

What must be decided politically, and with it metaphysically, is whether these abstractions and knowledges will be founded on deception and the reproduction of social inequality, or instead on some other equality. This decision, if we can correctly call it that, is not simply a decision made by individuals; it must be collectively discovered and constructed, and made into a necessity—just as the commodity has decided how one must survive under capitalism. All the while, it must also recognize the importance of confronting nonobjective forms of domination. Ultimately, it is through this process that statistics and machine learning present themselves as revolutionary objects: they call us to recognize, objectively, that they can no longer function under capitalist modes of production. The task of a revolutionary mathematics is, then, to discover and create another set of demands by which scientific objectivity necessitates a change of economy and production. In short, if science is to continue to feed us, both materially and intellectually, it can only do so against capitalism.

Chapter 8
Automated Abstractions and Alienation[1]

Whether forecasting gambling odds, the probability of rain or the future sales of Pop-Tarts, statistical models provide mathematical abstractions of the world; they take a varied and disparate world and reduce it to data that is then further reduced to a model that can extrapolate to new data. Yet unlike traditional Western thinking around abstractions, these models are not indexical of any stable world outside the constant flux of the data they ingest. The statistical abstractions that functionally underwrite the discoveries of machine learning rediscover their index in the fluidity of exchange, a foundation that does not point elsewhere to some transcendent ideal but rather immanently to the reasoned exchange of contracts and bets.

Within exchange's currents, these abstractions become increasingly mobile and modulatory; the same statistical processes can describe global social changes and individual shopping patterns. Like their more stable predecessors, these abstractions permit a freedom in dealing with the world, but at the cost of blinding us to their particulars. That is to say, a model is powerful in so much as

---

1 This chapter is coauthored with Cengiz Salman.

it allows one to understand and shape what has not been seen before, but it simultaneously risks overlooking what is most important. To account for the productive power and danger of these models, one must understand their abstractive force and the shifts in collective processes of abstraction they represent.

Freedom in Abstraction

Abstraction is freedom. As a word, "abstraction" etymologically connects to Latin's notions of "separation," "withdrawal" and "pulling away." As a concept, abstractions provide us distance from the muddy realities of the world, a detached epistemic position that allows things to meaningfully exist across different contexts: rain can become indicative of larger crop cycles or even evidence of global warming; individual events and data can be fit into larger patterns that can tell us about our world.

This distance from specificity has roots in the mythical realm: in Roman, Greek and Egyptian culture, deified abstractions, as expressed through the narrative lives of groups of gods and demigods, were treated as indices for specific concepts. Pax, the daughter of Jupiter and Justice, embodied peace. Thanatos embodied death, while Eros embodied love.[2] These abstractions were concepts personified, and vice versa; and each associated deity enjoyed concept-based cult worship and stature within the religious pantheon.

Abstractions like Pax would go on to lose their deistic distinction, dethroned from divinity while their conceptual namesakes remained in the heavens. These were the "forms," famously described by ancient Greeks like Plato in his Socratic dialogues, or the immaterial ideals by which the lived, physical world was understood as a mere shadow of their unchanging essence. For

---

2 Hasana Sharp, "Melancholy, Anxious, Ek-static Selves: Feminism between Eros and Thanatos," *Symposium* 11, No. 2 (2007): 315–331.

these idealist thinkers, true knowledge emanated from the irreducible, universal form, not from the specified, particular instance. In basic metaphysical terms, the particular was an inferior copy of the universal.[3]

While these abstract forms allowed a certain freedom for those who could wield them and were well represented by them, they necessarily effaced the experiences of others. The Enlightenment, colonialism, white supremacy, patriarchy and capitalism were all constructed atop abstractions that separate concepts from context, especially when it comes to the context of the marginalized. As philosopher and environmental activist Vandana Shiva writes, "Separability allows context-free abstraction of knowledge and creates criteria of validity based on alienation and nonparticipation, then projected as 'objectivity.'"[4] Even when abstractions threaten to cut away at authority, as we saw earlier with scientific

---

3 However, even these early deistic forms of abstraction bear some relation to exchange. French philosopher Jean-Pierre Vernant locates the origins of Greek abstraction in currency's capacity to standardize exchange. Vernant writes, "For the old image of wealth as *hybris* [i.e., hubris or insolence]—so charged with affective force and religious implications—legal tender substituted the abstract idea of *nomisa* [i.e., money], a social standard of value, a rational contrivance that allowed for a common measure of diverse realities, and thus equalized exchange as a social relationship." *The Origins of Greek Thought* (Ithaca, NY: Cornell University Press, 1982), 95. Like Vernant, Alfred Sohn-Rethel also recognizes that the real abstraction of money in the Greek polis established a generalized system of exchange, rendering qualitatively specific products of labor, including the labor of the enslaved, equivalent. More significantly, for Sohn-Rethel, the emergence of money in Greece served as the material a priori condition necessary for the construction of conceptual abstractions peculiar to its philosophical tradition: the one, the ideal, the many, and even the divine abstractions of gods like Pax (*Intellectual and Manual Labor*, 58–59). Consequently, money's functional capacity to exchange in Greece served as a constitutive factor in engendering the ideal abstractions specific to early Greek philosophical thought. Although the ancient Greek economy lacks the necessary features of a capitalist economy, e.g., generalized wage labor, Vernant and Sohn-Rethel clearly demonstrate how money in Greece prefigured capital's capacity to produce abstractions.

4 Vandana Shiva, *Staying Alive: Women, Ecology, and Development* (London: Zed Books, 1988), 22–23.

abstraction in the hands of freethinkers like James Jurin, the form of this expertise generated through detachment has an uncanny ability to reassert itself; after all, Jurin subsequently attempted to fall back on the authority of Isaac Newton. Abstractions tend to generate authority by empowering those who can (and are allowed to) make them productive.

Frankfurt school philosophers Max Horkheimer and Theodor Adorno argue that "the distance of subject from object, the presupposition of abstraction, is founded on the distance from things which the ruler attains by means of the ruled."[5] However, the once-hallowed epistemology of gods like Pax or the Christian divinity have been, over the course of the last centuries, replaced by an unencumbered, Western white masculinist mind. In the case of both God and man, knowledge and power is founded on the ability to abstract, to separate the subjective and objective world. And from the mind of this singular subject position, those who rule find power in seeing the world from afar, reclaiming the unreachable peak of Mount Olympus with the force of "objectivity."

Horkheimer and Adorno continue: "Under the leveling rule of abstraction, everything in nature [is] repeatable, and of industry, for which abstraction prepared the way, the liberated finally themselves become the 'herd' which Hegel identified as the outcome of enlightenment."[6] It is here that we can see the power of the commodity as an abstraction; it frees us at the same time it binds us to its calculations. Insofar as commodities conceal the social relations according to which they were produced by rendering qualitatively distinct forms of labor quantitatively equivalent through their exchange, commodities—as well as nature, industry, people, and knowledge—become "repeatable," objectified and hence countable. Such objectification treats the

---

5 Max Horkheimer and Theodor W. Adorno, *Dialectic of Enlightenment*, trans. Edmund Jephcott (Palo Alto, CA: Stanford University Press, 2002), 9.
6 Ibid., 9.

previously incommensurate as commensurate, thus rendering it understandable.[7]

To make nature repeatable—that is, to make life and goods exchangeable, and thus functionally traversable across different contexts—abstractions participate in what queer theorist Jasbir Puar calls the "productive tensions between abstraction and location." In essence, the particularity of location challenges the presumption of an abstraction's commensurability, of the universal's universality.[8] More specifically, to make something repeatable, and commensurate, is to forego any inquiry into these particulars—which could disrupt an abstraction and reveal what something was (or could be) outside the scope of its analytic use—while at the same time continuing to extract value from its concrete and local instantiation. In one sense, statistics and machine learning operationalize this productive tension; but, just like capitalism, they do so on the grounds of a fundamental commensurability. Machine learning's preference for correlations over causal explanations allows the inclusion of local data that may not immediately

---

7 As Ian Baucom, tracing the 1781 massacre of 132 enslaved Africans aboard the *Zong* (which was ordered by the ship's captain) and the ensuing court case over whether insurers were to pay the ship's owners, argues that even notions of justice and freedom are founded on ideas of exchangeability. He states: "The genius of insurance, the secret of *its* contribution to finance capitalism, is its insistence that the real test of something's value comes not at the moment it is made or exchanged but at the moment it is lost or destroyed ... In a money culture or an insurance culture value survives its objects, and in doing so does not just reward individual self-interest of the insured object's owner, but retrospectively confirms the system-wide conviction that the value was *always* autonomous from its object, *always* only a matter of agreement." Value under capitalism always points toward the ghost of a departed quantity, and what departs, or more accurately is made to depart, is invariably the most marginalized and excluded life. While perhaps there is a certain danger in any notion of exchangeability that must always be guarded against, finance capital seems to erect especially violent, inequitable and egregious forms of exchange. *Specters of the Atlantic: Finance Capital, Slavery, and the Philosophy of History* (Durham, NC: Duke University Press, 2005), 95.

8 Jasbir K. Puar, *The Right to Maim: Debility, Capacity, Disability* (Durham, NC: Duke University Press, 2017), xxv.

seem relevant to aid in the production of an abstract model. And yet, even these data must be selected, gathered and made digestible by an algorithm.

Thus, while maintaining the violence and domination of earlier forms, contrary to older modes of Enlightenment abstraction, the world of machine learning appears as an infinite chain of particulars, but also as an infinite chain of local and mobile abstractions—two views that, in the final analysis, amount to the same thing. All correlations are local only to the data on which they are trained. A model is never universal but always a particular and local abstraction. Machine learning appears to establish precisely what capitalism has always dreamed of: a smooth, universal lingua franca of epistemic and economic commensurability that always produces its truth only at the exact local moment of exchange. Algorithms allow the conversion of every bit of extractable data into interchangeable bits that can be compared and ultimately exchanged. Yet, these algorithms still produce and reproduce exactly the same kind of hierarchized difference that older forms of universality have.[9]

While abstraction, understood broadly, has long been an essential, if simultaneously productive and dangerous, component of human thought, machine learning mobilizes and liquefies these abstractions; machine learning algorithms, such as neural networks, allow computers to create abstractions that are only momentary correlations. Products are recommended to specific consumers based on one set of attributes this hour and then, as new data arrives,

---

9 Ramon Amaro describes the situation precisely in relation to face recognition: "To merely include a representational object into a computational milieu that has already positioned the white object as the prototypical characteristic catalyzes disruption on the level superficiality. From this view, the white object remains whole, while the object of difference is seen as alienated, fragmented, and lacking in comparison. From this perspective, the position of the black technical object (as Fanon and Wynter have thoroughly articulated) is lodged in a recurrent dialectic, where it attempts to valorize or recapture black life from within the confines of normalized logics while simultaneously desiring to disrupt its hold." "As If," *e-flux*, February 14, 2019.

on another set the following hour. Here, we can see the revolutionary implications of Bayesian analysis: frequentism sets up the experiment to determine repeatable, population-level abstractions, whereas Bayesianism allows the production of a nearly infinite field of hypotheses that can create an abstraction for each case. Bayesianism favors abstractions that produce a thought that is not universal, but merely momentary, whose only value is the value it fetches at market on a single day, or even for a single millisecond. In this way, it overflows and hollows out this productive tension, creating abstractions that are always local. It aims to banish all incommensurability in a sea of never-ending calculations.

A world without incommensurability is one that follows the utopian desires of Enlightenment idealists and now, increasingly, those of Silicon Valley engineers. This is a world imagined to be directly encoded into quantitative data, programmed and eventually automated—the "frictionless capitalism" of which Bill Gates dreamt in the 1990s. In short, it is a world without the traditional subject precisely because, in algorithmic form, the distinction between subject and object evaporates; one must act in accordance with what the algorithm will calculate about them, like a content producer constantly trying to optimize their search engine ranking. Each looks more and more like the other, and both subject and object become free, in a way, but only to follow the laws of the system. Thus, this freedom becomes its opposite; it is the freedom to choose what these object-abstractions demand of us. Corporations and researchers use the freedom of abstraction to run algorithms on seemingly disparate sets of data, conferring truth and value on billions of data records without knowing why there is a correlation; they know only that there is one, and that "objectively" they must choose to follow it. The only universal that remains is the structure of exchange itself and the constant fear of the Dutch book.

Google's PageRank algorithm offers a clear example of this dynamic. The algorithm is used to determine the "quality" of a web page based on the citation—or link graph—structure of the web, normalizing the quality of each page based on the quantity of "high-quality" websites that link to it. For PageRank, a web page's rank is not simply the total number of sites that link to it, but also the supposed quality (based solely on the calculations of the algorithm) of those pages that link to that web page. Consequently, Google may rank a site with fewer quality links higher than it does a site with more links from low-ranked sites.[10]

Google's computers, having crawled the majority of the World Wide Web, effectively simulate individuals surfing the web in order to determine the rates at which various pages will be visited. Statistically, we can understand PageRank as a measure of "the probability that the random surfer visits a page." And with a few additional "secret sauce" elements added to this simple but elegant formulation, Google's system functionally determines the "importance" of each website it has crawled. The PageRank of a single web page represents its importance. But a page's PageRank is wholly recursive: an important web page is one that is linked to from other important web pages. With PageRank, the entire concept of importance is founded in the algorithm and this recursion itself; so anyone who seeks to enter into this world must also model their own sense of importance on this algorithmic recursion.

To exist in this abstracted algorithmic world requires that individuals become open to, commensurate to and processable by the whole material infrastructure of algorithmic life. It is to live in the exclusive swell of these abstractions, tracing only how they are produced and how they are connected to each other. While this blow to the sovereignty of the subject may seem harsh, we must resist the urge to reassert ourselves as subjects in the Enlightenment sense, precisely because that subject remains ineffective and

---

10 Sergey Brin and Lawrence Page, "The Anatomy of a Large-Scale Hypertextual Web Search Engine," *Computer Networks and ISDN Systems* 30 (1998): 107–117.

encircled when algorithmic knowledge a priori reduces us to the supposedly immaculate, commensurate terrain of numbers and statistics.

Algorithmic Abstractions

Machine learning reduces everything to data in order to ground the world anew through abstractions, always searching for locally optimal solutions in real time. To offer another example, take Google's image recognition system, a "deep learning" approach, which means in essence that there are multiple layers to the neural net that learn different levels of abstraction in order to allow researchers to train a network of computers to classify images. Through this network, computer scientists at Google were able to create the category of "cat" according to the statistical commonalities from 10 million YouTube video screen grabs.

"We never told it during the training, 'this is a cat,'" said Google fellow and researcher Jeff Dean. "It basically invented the concept of a cat."[11] Google's network architecture, composed of 16,000 interlinked core processors, had mimicked the firing of individual neurons, functionally enabling their computers to recognize common objects within large sets of data—all without labeling and without preconditioning. In doing so, Google's algorithm invented "cat" without having any notion of what a cat was. There is no construction of cats through the universal a priori of cat. Nor is there any group of priests or experts defining a universal cat through some presumed, particular position of distanced objectivity.

Instead, Google's cat is made from the ground up, built atop pattern recognition algorithms. These algorithms search for local

---

11 John Markoff, "In a Big Network of Computers, Evidence of Machine Learning," *New York Times*, June 25, 2012.

features and combine these into more general abstractions.[12] This freedom of abstraction begins with pixels, which become lines, which become shapes—each step forming another hidden layer in Google's artificial neural network. Then those shapes are aggregated at the next level according to graphic similarities (what colors and shapes are most often present). Next, a barrage of millions of random image thumbnails are thrown at Google's 16,000 computer processors so that the visual component of what we now call "cat," and what a computer might know only as a series of polygon "edges" and commonality measures, can be made and discursively massaged so as to be called a cat.

It should be remembered that while these abstractions appear to be built ex nihilo, the training data, which comes from YouTube videos, is the by-product of a complex, biased and mediated process of social production. Thus, from this mass of socially produced data, an automatic abstraction is reflected back to us as its truth and objective product. In a way, this is analogous to the construction of value: an ongoing process of acquiring or creating data is constantly put into exchange with itself, working on itself over and over until it can be re-presented as truth.

This is a key consequence of the move to algorithmic knowledge: what Google says is (and, conversely, what it says is not) a cat is based on socially produced data, and the singular abstraction of Google's algorithm. Even knowing exactly how the algorithm works, when it runs in real time we have no access to the middle-level abstractions it produces. The computational opacity that created this cat is akin to quantum physicist Erwin Schrödinger's famous feline, which as a result of random radioactive decay was said to be simultaneously alive or dead: we have no idea if what is being processed is in fact alive, dead or even an actual cat. Moreover, while in the end we may be concerned mostly with the

---

12 Quoc V. Le, Marc'Aurelio Ranzato, Rajat Monga et al., "Building High-Level Features Using Large Scale Unsupervised Learning," in *Proceedings of the 29th International Conference on Machine Learning* (2012).

adjudication of "cat" or "not cat," just like Schrödinger's cat, the understanding of these judgments are ultimately probabilistic: the algorithm is not designed to always know what is and is not a cat, but rather to manage an uncertain world and guess correctly often enough. In a world made of data (as Google's cat is nothing but strong, sequential correlations between statistically similar objects), meaning is a matter of algorithmic output: a probable cat (see the image below).

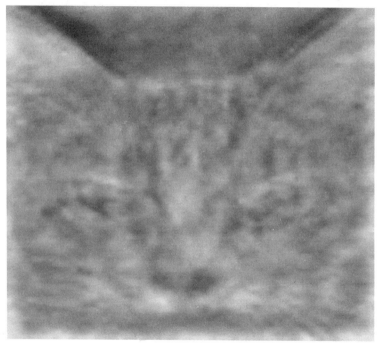

The Google algorithm's archetypal image of a cat constructed from YouTube stills.

Accordingly, we see how the world gets dynamically objectified before our very eyes. The vocabulary of this algorithmic Adam becomes the dynamic, emergent expressions that regulate the knowledges of flora and fauna—and indeed, any other concept the algorithm learns. But unlike the original Adam, whose name carries with it the imprimatur of God, these names are always particular, local and fluid; in fact, they do not actually exist, for the algorithm merely assembles a composite and stores it somewhere in memory. While we may look at this objectified output and shake our heads in agreement that it has discovered a cat, the endgame of Google's algorithm is to repeat this process millions and millions of times. Ultimately, the abstractions an algorithm uncovers create a reality that must be reckoned with whether the abstractions themselves are "true" or not.

For example, an algorithm used by the state of Michigan to determine whether unemployment insurance claims were fraudulent falsely accused over 34,000 people of fraud. These people saw their unemployment payments automatically cut off. While these errors are finally being corrected, those thousands of individuals still had to deal with the assessments of this software; since the algorithm categorized them as fraudulent, they functionally were, regardless of their actual behavior.[13] Especially because the state and the market, under their expansive neoliberal logic, function in a predatory manner, these systems serve less to predict the future than to define it and provide an authoritative force for that definition.[14] Such algorithms constantly define and redefine what the abstracted knowledge of the world means, but on a ground that is always the product of capitalism and of the injustices of history that have constructed our current moment.

Enclosing the General Intellect

As machine-learned abstractions produce knowledge like Google's cat, the instructions and parameters that assembled that knowledge become wholly classified, and any who may hope to fix the

---

[13] Robert N. Charette, "Michigan's MiDAS Unemployment System: Algorithm Alchemy Created Lead, Not Gold," *IEEE Spectrum*, January 24, 2018.

[14] Jackie Wang, *Carceral Capitalism* (Cambridge: MIT Press, 2018), 16–17, 43–48.

errors generated by these algorithms are always running behind their automated production. In this way, digital capitalism utilizes this fluidity to enclose knowledge itself.[15]

Knowledge founded on an external referent—whether God or some heroic theory of scientific truth—functionally resists enclosure, for this ground must be shared among society to make the knowledge socially usable. For Christianity to become socially productive, it had to ally with the state and abandon its status as a small cult. Of course, states can and do build privatized legal frameworks (such as patents) atop this shared ground to allow the protection of corporations' market positions, but knowledge of the world itself still belongs to the community that shares this ground.

However, once knowledge finds its ground exclusively lodged in the mechanisms of capitalist exchange, it exposes itself to enclosure—analogous to the process by which, from the thirteenth century on, shared grazing land in Britain was enclosed and made into private farmland. In the history of statistics, we see a similar confining process in the unmooring of probabilistic knowledge from a shared (if selfish) research community that begins with Jerzy Neyman and Egon Pearson and finds its ultimate expression in Leonard Savage and Bruno de Finetti. And now, with the contemporary operationalization of statistics in modern machine learning, the process of enclosing knowledge has reached an intensity and extent previously impossible. For proof of this drive for enclosure, one merely has to look to the present clamor and price for large proprietary datasets, or the seemingly nonstop revelations that corporations like Facebook have provided secret access to user data to large corporate partners.[16]

---

15 This mobile, ever-shifting epistemic terrain is structurally aligned with that theorized by Gilles Deleuze and Zygmut Bauman. Zygmunt Bauman, *Liquid Modernity* (Malden, MA: Polity, 2000); Deleuze, "Postscript on the Societies of Control," *October* 59 (1992): 6.

16 Gabriel J.X. Dance, Michael LaForgia and Nicholas Confessore, "As Facebook Raised a Privacy Wall, It Carved an Opening for Tech Giants," *New York Times*, December 18, 2018.

Behind these bulwarks of privatized, abstracted knowledge reside the generative spirits of what Marx called "primitive accumulation"—the origin processes by which accumulated surplus wealth could become capital.[17] Yet such initial accumulations of surplus, in opposition to the ideal meritocracy of classical thinkers like Adam Smith, must reside outside the formal laws of capitalism. Historically, it was violence, colonialism, theft and enslavement that expropriated and impoverished others in order to create the massive stores of wealth needed to build capitalism's foundations and maintain their function. This "systematic theft of communal property," writes Marxist geographer David Harvey, began with a "grand movement of enclosure of the commons," a dual process of both colonial exploitation and the privatization of previously public goods.[18]

As capitalism continues its turn toward information as a new frontier, we enter a second enclosure movement that privatizes not land but ideas—ideas once considered the common property of society and culture writ large.[19] From the proprietary computer code that undergirds much of the world's digital infrastructure to profit-seeking scientific research on the human genome, patents on intangible knowledge have become engines of information capital.[20]

---

17 Karl Marx, *Capital*, Vol. 1 (London: Penguin Books, 1990), 873–940.

18 David Harvey, *A Companion to Marx's Capital* (London: Penguin Books, 1990), 295.

19 James Boyle, "The Second Enclosure Movement," *Renewal: A Journal of Labour Politics* 15, No. 4 (2007): 17–24.

20 While some economists believe that the legal enforcement of intellectual property rights spurs innovation and economic growth, scholars critical of such privatization have powerfully argued that patents both reproduce existing inequalities by concentrating wealth and knowledge in the hands of a privileged class of capitalists, and limit the capacities of others (particularly those in the developing world) from accessing and applying such information for social, cultural and economic benefit. See Ronald V. Bettig, *Copyrighting Culture: The Political Economy of Intellectual Property* (Boulder, CO: Westview Press, 1996); C. Ford Runge and Edi Defrancesco, "Exclusion, Inclusion, and Enclosure: Historical Commons and Modern Intellectual Property," *World Development* 34, No. 10 (2006): 1713–1727.

These forces seek not to generate wealth so much as expropriate, steal and privatize wealth that exists elsewhere.

A new kind of primitive accumulation arises from the specifics of this second enclosure. Here, the privatization of once-common agricultural lands is replaced with the privatization of once-common social knowledge and conditions of social life itself, or what Marx calls the general intellect.[21] This most recent enclosure is carried out by many long-existing legal implements: patents, trade secrets, nondisclosure agreements, and so on.[22] Moreover, the establishment of massive global supply chains owned by multinational corporations allows the expropriation of labor power and wealth from the global South. This further facilitates the movement of an increasing number of workers into urban areas, accentuated simultaneously by the privatization of lands by agribusiness and the possibility of higher wages in urban areas, further increasing the global supply of labor and its reserves.[23] While these methods have long been part of capitalism, they are now turned directly into the production of surplus value through the automatic and algorithmic production of knowledge, which is accompanied by rising global demand for extraction and production.

This metamorphosis is exemplified by the case of high-frequency algorithmic trading and the intellectual property regimes that facilitate gene patenting. The former acts essentially as a corporate tax on slower traders, while the latter privatizes the intellectually and socially productive power of information that exists out in the world, often through the theft of intellectual property owned by marginalized communities.[24]

---

21 Karl Marx, *Grundrisse: Foundations of the Critique of Political Economy* (London: Penguin, 2005, repr.), 706.

22 Ishmael Burdeau, "The Last Great Enclosure: The Crisis of the General Intellect," *Journal of Labor and Society* 18, No. 4 (2015): 649–663.

23 Farshad Araghi, "The Great Global Enclosure of Our Times," in *Hungry for Profit*, Fred Magdoff, John Bellamy Foster and Frederick M. Buttel, eds. (New York: Monthly Review Press, 2000), 145–160; Midnight Notes, "The New Enclosures," *Midnight Notes* 10 (1990): 1–100.

24 Vandana Shiva, *Biopiracy: The Plunder of Nature and Knowledge* (Berkeley: North Atlantic Books, 2016).

Where this privatization becomes especially clear is in the growth of what economic theorist Nick Srnicek outlines under the name "platform capitalism," where companies own the software and hardware on which social and economic interactions take place.[25] As famously recounted by both anti-capitalists and cyberutopians alike, while Uber owns no real taxis, Facebook has no real friends, and Airbnb owns no properties, all three companies make billions in revenue solely by enabling the most basic of market interactions. The new digital economy has overseen the extraordinary proliferation of platforms—from Uber to Etsy to Amazon—and even traditional manufacturing companies like General Electric and John Deere have attempted to transform themselves into proprietary data companies. John Deere, for instance, has in essence created a farming platform where farmers are unable to repair their own equipment, and where the data they produce is aggregated by the company.[26]

Srnicek argues that these platforms generate their profits through two interrelated methods. First, they supply an infrastructure for exchange that benefits from network effects: when enough sellers and buyers are using the platform, both groups are further attracted to the platform because their opposite is already there. These effects make commercial competition incredibly difficult, since any upstart platform will lack the requisite numbers of buyers/sellers or products to choose from. They are then able to extract a percentage from the transaction, essentially requiring those on the platform to pay rent. Second, and most important for our purposes, platforms extract data from users and their transactions. This data, abstracted by machine learning algorithms, allows for more rapid and effective matching of buyers to sellers within the platform, as seen in Uber's ability to outpace traditional taxis in getting drivers to passengers. This virtuous cycle enhances an

---

25 Nick Srnicek, *Platform Capitalism* (New York: Polity, 2017).
26 Rian Wanstreet, "America's Farmers Are Becoming Prisoners to Agriculture's Technological Revolution," *Motherboard*, March 8, 2018.

established platform's competitive advantages, further accelerating the hoarding of capital and market position.

Like algorithmic alchemists, these platforms have found a way to extract value from exchange twice. For not only do they tax an exchange (e.g., Uber takes a percentage of each fare, while Facebook charges for advertising), but they then extract data from that transaction, algorithmically processing it to make the platform's matching system even more efficient. The data can also be sold to other companies, allowing its further transformation into profit. Now not only the spaces, but the very means by which people shop, communicate and travel are being directly privatized. The once-public (or at least semi-public) cacophony of the market itself is now being enclosed, and with it, any public knowledge that could be learned from exchange.

The process of enclosing the general intellect begins with what the Marxist philosopher Paolo Virno calls "mass intellectuality," to which, "the entirety of post-Fordist living labour [belongs] ... to the extent that it is the depository of cognitive competencies that cannot be objectified in machinery. Mass intellectuality is the prominent form in which the general intellect is manifest today. What is at stake is obviously not the scientific erudition of the individual labourer."[27] We should take Virno's statement here at its most literal: there exists an unbridgeable metaphysical gap between the general intellect and the scientific production of the individual laborer. For science or computation to be productive, it must be productive directly at the level of the general intellect. Ronald Fisher's insistence that the individual scientist is the one who should know can never be directly productive. While this process of enclosure may appear profitable for a time, it directly threatens its own foundations, and scientific knowledge production along with it.

While some autonomist Marxists have read the development of the general intellect as a force that would necessarily diminish

---

27 Paolo Virno, "General Intellect," *Historical Materialism* 15, No. 3 (2007): 6.

labor time completely, offering a leftist promise of techno-utopianism, the possibility of its enclosure and privatization suggests, rather, that it is a site of capitalist contradiction—albeit one whose resolution in the favor of labor is not guaranteed.[28] In the words of information and media scholar Nick Dyer-Witheford, it is valuable to see "the issue within an antagonistic perspective, seeing the noosphere contested in class war—and this is the prospect the theory of 'general intellect' opens."[29] It opens up the space for the development and future of humanities collective potential, but offers no promise of a specific development.

On a metaphysical level, the withdrawal and collapse of a transcendental anchor for knowledge simultaneously allows for its privatization (since this transcendent other no longer guarantees its possession by a community often founded on its own exclusions) and demands its collectivization (since it requires the collective act of exchange based on shared knowledge for its meaningful production). While there should be no fundamental opposition to individual pursuits of knowledge, Fisher's "selfish and perhaps heretical aim of understanding for oneself the scientific situation" should be of no concern to those who hope to make science productive for the general benefit.[30]

Indeed, for the general intellect to function, it must remain open and social; otherwise, dissimulation will take hold, and, as we witness with the litany of recent defeat devices and the privatization of platforms, rent-seeking behavior will necessarily win out over any socially beneficial abstraction of the world. Here, we encounter the primary contradiction of knowledge in contemporary digital capitalism: for statistics and machine learning to be

---

28 This reading is especially present in Antonio Negri and others' writing in *Futur Antérieur*. For an excellent overview see Nick Dyer-Witheford, "Cyber-Negri: General Intellect and Immaterial Labor," in *The Philosophy of Antonio Negri*, Timothy Murphy and Abdul-Karim Mustapha, eds. (London: Pluto Press, 2005).

29 Ibid., 143.

30 Fisher, "Statistical Methods and Scientific Induction," 70.

productive, they must function at the level of the general intellect, not resigned to the private enclaves of proprietary datasets. But, the more successful datasets are in generating efficient abstractions for profit, the greater incentive there is to privatize, enclose or simply fake them. Once the production of data and its abstractions are enclosed within Facebook's or Uber's databases, those abstractions invariably end up elucidating not some ideal, general will but what we might call the "enclosed intellect"—one that can produce only knowledge from a specific commodified position according to each firm's location in the market, where that knowledge is determined exclusively by its ability to produce a profit. This whole process invariably replicates and objectifies all of the social biases and exploitations that exist there. Once this happens, then non-knowledge, dissimulation and defeat devices become more exchangeable, and hence profitable, than knowledge or any aspiration for a more just world.[31] We can look, for example, to Theranos, the company that falsely claimed to have developed a new more efficient technology for blood testing, raising hundreds of millions of dollars on such fake knowledge.[32] Under late capitalism, completely fake abstractions can be just as valuable as ones that correspond closely to the world.

The Capitalist Calculation Problem

The hyper-capitalist, privatized enclosure of the general intellect reflects an inversion of liberal economists Ludwig von Mises and Friedrich Hayek's famous critique of socialism. For von Mises, and later Hayek, socialism suffers from what they call a "calculation problem."[33] This problem, for them, is that a centralized, bureaucratic

---

31 Jathan Sadowski, "Potemkin AI," *Real Life*, August 6, 2018.
32 Nick Bilton, "How Elizabeth Holmes's House of Cards Came Tumbling Down," *Vanity Fair*, October 2016.
33 Many have convincingly critiqued Hayek's position since his writing. See David Harvey, *A Brief History of Neoliberalism* (Oxford: Oxford University Press,

entity is unable to properly ascertain the optimal distribution of goods. How would a Chilean state planner in Santiago know how many mobile phones should be sent to a store in Viña del Mar?

Von Mises and Hayek argue that markets solve this problem by using price to calculate a rational, and efficient, distribution of goods through supply and demand. If the market is allowed to run its course, the subjective interests of all consumers, they reason, will become accurately reflected within the infinite cycle of exchange. Thus, the laws of supply and demand allow the market to move goods where they are most needed. If we are to believe these two, capitalism bridges the metaphysical gap separating individual knowledge from the general intellect, or the particular from the universal, by creating an imaginary world where each individual's private calculations can create a stable economy. This resolution conceives of individuals as distributed computers who together buy and sell their labor and goods so as to calculate the "optimal" distribution of goods and labor. In essence, capitalism produces the same mobile, fluid, real-time abstractions that algorithmic systems are supposed to achieve.

Hayek, who arguably founded the political economic framework we now call neoliberalism, laid out the importance of this calculative process in his 1936 address to the London Economic Club:

> Economics has come nearer than any other social science to an answer to that central question of all social sciences, how the

---

2005); Jamie Peck, *Constructions of Neoliberal Reason* (Oxford: Oxford University Press, 2010); Dieter Plehwe, "Introduction," in *The Road to Mont Pèlerin: The Making of the Neoliberal Thought Collective*, Philip Mirowski and Dieter Plehwe, eds. (Cambridge, MA: Harvard University Press, 2009); Rob Van Horn and Philip Mirowski, "The Rise of the Chicago School of Economics and the Birth of Neoliberalism," in *The Road from Mont Pèlerin*; and Philip Mirowski, *Never Let a Serious Crisis Go to Waste: How Neoliberalism Survived the Financial Meltdown* (London and New York: Verso, 2013).

combination of fragments of knowledge existing in different minds can bring about results which, if they were to be brought about deliberately, would require a knowledge on the part of the directing mind which no single person can possess.[34]

For both von Mises and Hayek, this process is only possible under capitalist market conditions, because no bureaucracy can compute a "rational" economic distribution that fully takes others' individual knowledges into account. But as information and data become more integral to the economy, it is now capitalism that becomes wholly unable to calculate a rational distribution (even by capitalism's very inequitable definition of "rational").

This point is exemplified by the perpetual state of crisis that defines contemporary global capitalism. For example, on September 16, 2013, the US Federal Reserve announced that it would continue its post-2008 financial stimulus strategy of purchasing government bonds. The decision, announced at the Reserve's headquarters in Washington, DC, was made at exactly 2 p.m. (as measured by the national atomic clock). A mere three milliseconds after this information was announced, several large asset orders were placed in Chicago.

At its fastest, given the speed of light, which also limits the speed at which information can travel, it should take such news at least seven milliseconds to make the fiber-optic journey from DC to Chicago. While reporters were told about the decision prior to the announcement, they were sequestered in a room, unable to communicate until 2 p.m. In the high-stakes world of high-frequency trading—where computers attempt to make massive profits off of millisecond advantages in trading speed—these four milliseconds of advance notice were worth substantial amounts of money: an estimated $600 million changed hands before other

---

[34] Friedrich Hayek, "Economics and Knowledge," in *L.S.E. Essays on Cost*, James M. Buchanan and George F. Thirlby, eds. (New York: New York University Press, 1973), 66.

Chicago traders were aware of the Federal Reserve's decision. Someone had stolen these incredibly valuable milliseconds, acting on this information in order to trade hundreds of millions of dollars before anyone else could act.

High-speed financial transactions come as a consequence of what David Harvey argues has become central to our contemporary economic structure: the compression of space-time made possible by the rise of high-speed networked digital capitalism.[35] Since the 1970s, accelerating developments and mobilizations of communication technologies have intensified the circulation and consumption of immaterial and ephemeral commodities, from memes to financial derivatives.[36] These informational commodities no longer simply manage material production, but are directly productive themselves. In the labor processes of a thoroughly virtualized and informationalized post-Fordist economy, Paolo Virno argues, "thoughts and discourses function in themselves as productive 'machines' ... in contemporary labour and do not need to take on a mechanical body or an electric soul."[37]

Knowledge in the knowledge economy is not simply congealed in, and mobilized by, sophisticated machines. Rather, knowledge itself has become a quasi-independent productive force, immanent to the social relations of contemporary capitalism: knowledge no longer increases the speed by which industry can produce goods but, as it becomes a force of automation, produces value on its own (e.g., Bitcoin mining, high-frequency trading, automated news reporting, etc.). Production is no longer localized in factories. Capitalism subsumes and transmutes all social relations into modes of production such that society itself becomes a "social

---

35 David Harvey, *The Condition of Postmodernity: An Enquiry into the Origins of Cultural Change* (Cambridge, MA: Blackwell, 1992).

36 Maurizio Lazzarato, "Immaterial Labor," in *Radical Thought in Italy: A Potential Politics*, Paolo Virno and Michael Hardt, eds. (Minneapolis: University of Minnesota Press, 1996), 133–146; Ivan Ascher, *Portfolio Society: On the Capitalist Mode of Prediction* (Cambridge: MIT Press, 2016).

37 Virno, "General Intellect," 5.

factory" where all social interactions become preludes to production, now exemplified by constant unpaid cultural production and the gig economy, where one is always networking.[38]

We are witnessing the rise of an economy that places immense importance on the aggregation of minuscule fragments of information: search queries, GPS locations, Facebook posts, to name a few. While labor time was surely spent figuring out how to steal these four milliseconds, the material economic payoff was made from pure knowledge and what amounts to a form of theft: knowledge of the exchange value of financial instruments before other traders. Thus, it is possible to offer an alternative definition of the so-called "knowledge economy." This knowledge economy is characterized not by the productive use of knowledge, but rather by its exact opposite: the unproductive theft and privatization of knowledge, and hence the theft and privatization of the general intellect. Rather than these three stolen milliseconds being used to advance some form of general knowledge, they functioned to valorize capitalism by impoverishing collective knowledge.

In this way, current uses of statistical inference and machine learning merely discover the rules of the game—not how to change them. As long as the stakes are built around profit, statistical methods can only be a science of how best to tilt the scales or cook the books. Thus, the question of statistical inference and machine learning is a priori a question of economy, social production and knowledge: it cannot stand outside the political economy that underwrites it.[39]

When economic value production is shaped, or at the very least facilitated, by personal degrees of uncertainty—as the Bayesian revolution places the burden of knowing on the individuals who exchange, whether humans or computers—there becomes an

---

38 Antonio Negri, *The Politics of Subversion: A Manifesto for the Twenty-First Century* (Cambridge, UK: Polity, 1989).

39 Malcolm Harris, "Glitch Capitalism: How Cheating AIs Explain Our Glitchy Society," *New York Magazine*, April 23, 2018.

increasing economic incentive to lie: under contemporary capitalism, corporations now fabricate their firms' economic conditions with a regularity and scale that was once seen to be the exclusive preserve of Soviet bureaucracy. For Hayek, the price mechanism works to distribute knowledge throughout an economy because it requires that individuals share knowledge in order to evaluate the price of a commodity, but an increasingly fluid and privatized economy makes dissimulation the rule.[40]

Through this privatization we see the creation of nearly insurmountable roadblocks to scientific and technological discovery outside the confines of available data and the drive for profit.[41] We need only to look at the litany of inaccurate algorithmic systems; the replication crises in the sciences; the scandals over user data and political advertisements that seem to elicit only muted apologies from companies like Facebook; or Kobe Steel's falsification of quality control data for their metals for over a decade, which potentially compromised the structural integrity of cars, airplanes and trains.[42]

Like financial crises and underemployment, these duplicities are not aberrations from an otherwise well-functioning economic system. Rather, they represent what can be called a "capitalist calculation problem."[43] Probabilistic knowledge requires that as much data as possible be arranged and calculated in a single place. But under the reign of probability, markets discourage the dissipation of data and instead horde as much as possible. Without some other sort of system, some other mode of calculation—in essence some other means of exchange than capitalism—probabilistic

---

40 Evgeny Morozov, "Digital Socialism?," *New Left Review* 116/117 (March–June 2019).

41 Nick Srnicek and Alex Williams, *Inventing the Future: Postcapitalism and a World without Work* (London and New York: Verso, 2016).

42 Robin Harding, "Kobe Steel Admits It Falsified Data on Aluminum and Copper Parts," *Financial Times*, October 8, 2017.

43 Richard Seymour, "Marxism, the Bourgeoisie and Capitalist Imperialism," *Lenin's Tomb* (blog), April 30, 2006.

knowledge production will continue to favor cheating and dissimulation. As we saw in Chapter 4 with Jonah and his shipmates, lots tell us nothing about an event without the epistemic support of a larger theological and cleromantic system. But under capitalism, this support system increasingly turns against the production of *usable* knowledge in favor of *valuable* knowledge, which more often than not means the production of socially useful knowledge is replaced with an arms race that aims at the impoverishment of the knowledge of one's interlocutors. In the final analysis, what is objectified through statistics and machine learning is precisely and exclusively the demands of the market. These systems serve only to re-present both the concrete and abstract domination of contemporary capitalism in objectified form.

The enclosure of the general intellect forecloses the possibility of building collective knowledge. Within a firm's proprietary enclosure, statistics is only able to produce abstractions according to available, local, and parceled-off data and knowledge. In doing so, machine learning ultimately serves only the purpose of what is sometimes called "local optimization"—of finding a solution that is best for a given part of a problem. This is not to say there is no importance to local, situational and specific knowledge—in fact this is the only meaningful place knowledge can come from—but as long as the specificity of the local is determined exclusively by profit, it can only be a reflection of global systems of expropriation and exploitation.

Naturality and Ontology

Artificial intelligence is still only able to solve the problems that are given to it.[44] While computers can process increasingly large datasets, these datasets contain only a fraction of all the attributes

---

44 See Yuk Hui, *On the Existence of Digital Objects* (Minneapolis: University of Minnesota Press, 2016).

that could affect the possible outcomes. Moreover, even in the most unsupervised of machine learning problems, an engineer must on some level determine what a problem is and what a successful solution to such a problem may look like. The freedom in abstraction is a freedom conditioned by this definition. To abstract is to provide distance from a certain, specific subjective position, all while basing the entirety of that abstraction within the knowledge afforded to that position.

And as proprietary, opaque systems continue their ascent toward a persistent, real-time production of abstractions, the force of those abstractions' influence finds little effective response in any political theory that requires stable terrain. The very concepts through which the world is now defined are no longer fixed, and thus they have become even more inaccessible. When Google CEO Sundar Pichai testified before Congress in 2018, conservative lawmakers repeatedly accused the search engine of returning liberally biased results. Pichai was largely able to evade these accusations by explaining methodological issues rather than defending specific results.[45] Indeed, the force and authority of these calculations cease to be localizable to the company or its employees, appearing instead to come from some method that must be followed. While old forms of abstraction alienated all but an elite aristocracy from the world, new computational ones alienate all humanity from thought itself. One cannot know, and thus politically act, if the subject has an inadequate understanding of both the terrain on which they stand and the nonlinear consequences of their actions.[46] While politics has always been confronted with nonlinear dynamics, the proliferation of systems that attempt to automatically modulate these

---

45 House Judiciary Committee, "Hearing on Transparency and Accountability: Examining Google and Its Data Collection, Use and Filtering Practices," 115th Congress, December 11, 2018.

46 Antoinette Rouvroy, "The End(s) of Critique: Data Behaviorism versus Due Process," in *Privacy, Due Process and the Computational Turn: The Philosophy of Law Meets the Philosophy of Technology*, Mireille Hildebrandt and Katja de Vries, eds. (New York: Routledge, 2013), 143–168.

dynamics foreclose any hope of a politics founded on a subject that would simply choose differently.

As thought and knowledge become increasingly alien and unintelligible in any traditional and immediate sense, any recourse to a politics that aspires to an unalienated, "natural" subject—or even centers a subject conditioned only by the objective, and observable, stakes of the world—will eventually fail, for it will inevitably be unable to account for the material and political force these algorithmic and statistical systems create. This is precisely because such a subject position presupposes a transcendental perspective that opaque machine learning algorithms do not afford. That is to say, this subject would need to be endowed with a capacity to stand outside or above the situation—an impossible omniscience and transcendence—in order to know how these technologies mediate social relations through the production of abstractions. Moreover, as philosopher and cultural theorist Sylvia Wynter has demonstrated, the very concept of the human ties together its earlier theological and transcendental force with a naturality that never overthrows the hierarchized understanding of existence but instead encodes it into a notion of humanity that always excludes those who are not male, white and so on. It is only by abandoning this commitment to the natural and the true that these systems of valuation can be overturned.[47]

The subject, then, must be thought not as some veritable, emanating truth. One's subjectivity is possessed wholesale by the universality of the system, and vice versa; in short, non-locatable objectivity twists into subjective particularity. In this trapping, one can believe—or identify—however one wants, but one must necessarily follow the laws of probability that have been programmed into the system. The laws of probability, like the market upon which they are founded, function with or without the belief of

---

47 Sylvia Wynter, "Unsettling the Coloniality of Being/Power/Truth/Freedom: Towards the Human, after Man, Its Overrepresentation—An Argument," *New Centennial Review* 3, No. 3 (2003): 257–337.

those who bring their knowledge to market. So, the subject is free, but only in so much as they can follow the law of the system governed by what is made to count, which under capitalism is always a question of what counts at market. The particular is thus alienated in the universal, and the universal is simultaneously alienated in the particular.

When belief, experience and subjectivity in general are conditioned by the constraints of exchange, a property owner's declaration that it is not they who are racist, but rather the market, reflects precisely the logic by which the system of racial capitalism is made to function.[48] Of course such claims are racist, but the subject who makes them believes they come from somewhere else, namely the objectivity of the market. As structural white supremacy is objectified into the statistical matrix of one's property value, life gets trapped within the abstractions of calculating process—an epistemic paralysis that resists critique. Thus, we witness the continuation of racist housing policies in algorithms' preferential treatment of white mortgage applications since the dataset upon which such calculations are based incorporates a history of racism in housing and lending.[49]

---

48 Practices of redlining in Michigan's major cities involved the systematic denial of mortgage financing based on race, geography, or other noneconomic criteria until such practices were de jure abolished with state legislation in 1978. For an explanation of the law, see J. Richard Johnson, "Michigan's Redlining Law," *Detroit College of Law Review 1978*, No. 4 (Winter 1978): 599–624. For a history of redlining and white flight in Detroit, see Thomas J. Sugrue, *The Origins of the Urban Crisis: Race and Inequality in Postwar Detroit* (Princeton, NJ: Princeton University Press, 2005). Practices of redlining remain de facto prevalent in Detroit and Lansing, Michigan, in 2018. The president of the Michigan Mortgage Lenders Association recently suggested that the automation of lending makes it difficult to recognize practices of discrimination. See Aaron Glantz and Emmanuel Martinez, "Detroit-Area Blacks Twice as Likely to Be Denied Home Loans," *Detroit News*, February 15, 2018. For more information about contemporary practices of redlining in Michigan and the US more broadly, see Aaron Glantz and Emmanuel Martinez, "For People of Color, Banks Are Shutting the Door to Homeownership," *Reveal*, February 15, 2018.

49 Jordan Pearson, "AI Could Ressurect a Racist Housing Policy," *Motherboard*, February 2, 2017.

The particulars of the subject's lived, subjective experience become wrapped up with the structures of these computationally produced abstractions of neoliberal capitalism. In metaphysical terms, statistics have become the formal science—now governed by the market itself, thanks to the Bayesian revolution—by which the relation of the particular gets objectively connected to the universal. The historical conflict between one's life and the structures around that life have not been reconciled, but merely reframed into mobile, fluid and market-based relations. Some are granted more or less freedom to abstract, but still these abstractions are mediated by the demands of capitalism.

In order to break this metaphysical stranglehold, we cannot seek a return to some natural, pretechnological, prestatistical, precapitalist form. Nor can we take seriously the resolute position of the liberal subject—of one who can know, and functionally audit, the complexity of these algorithmic systems so as to solve these problems through reform alone. Instead, we must transform the necessity of what is already conventional and structurally fluent within the statistically ordained world: abstraction in all its alienatory power.

So, the act of abstraction itself—and even and especially its mobilization against some fixed transcendent reference point—does not necessarily have to be a force for exploitation. While these abstractions necessarily elide some attributes of the world they model, they are still potentially productive.[50] But if they are to be socially productive, they—along with the technologies and techniques that produce them—must be disaggregated from their capitalist ruin. While the

---

50 In a blog post, philosopher Adam Kotsko recognizes Scott Ferguson and Anna Kornbluh as respectively conducting research that conceptualizes the liberatory potential of abstractions. See Adam Kotsko, "Reading Agamben with Ferguson," *Provocations* 2 (2018). While Kornbluh's work on the topic is unpublished as of this writing, Scott Ferguson draws on the resources of modern monetary theory (MMT) to argue that money, if freed from the parochial concerns of reactionary politicians, could function as an unconstrained and infinite resource for addressing many contemporary social issues while also constructing a broader public. *Declarations of Dependence: Money, Aesthetics, and the Politics of Care* (Lincoln, NE: University of Nebraska Press, 2018).

steps that must be taken are far from clear, collectively we must turn to the work of a revolutionary mathematics to value different mysteries, different relations between universals and particulars, and new forms of abstraction that are freed from the production of privatized knowledge and value. The old mysteries of God, the state and now even capital lay in disrepair and are incapable of supporting the production of knowledge and our collective existence. This is not to say that they were ever able to produce, for large portions of humanity, much more than ruins; but now even their utopian, scientifically driven metaphysics are directly reduced to ruin.

In sum, we must theoretically and practically engage the production of knowledge, imagining different forms and different computations. But let us be clear: this does not and cannot mean a simple fetishism of the future for its own sake. We cannot be so hopeful as to assume that the force of these contradictions promises any respite. At best, they provide an opening, an opportunity to reconfigure the very force of objectification and the production of knowledge and value.

As the ability to record and process data continues to increase exponentially, it will be on the level of the general intellect that we may be able to take advantage of these technologies. In short, only through their communal, socialized usage will we be able to turn data into meaningful and productive knowledge. Today, it is capitalism that suffers from a calculation problem, and only some other form of exchange and abstraction will be able to calculate a more meaningfully rational and just usage of goods and labor.

Alienation

To fully accept the productive weight of twenty-first-century science and digital technologies requires a reconsideration of the Marxist theory of alienation.[51] Marx prominently argues that the

---

51 Srnicek and Williams, *Inventing the Future*, 82.

process of capitalist production is a process of the alienation of labor. Artisanal workers and early farmers produced the very goods they either consumed or whose sale they controlled. While for many workers (especially the enslaved and indentured) precapitalist production alienated labor, capitalism generalized this alienation in the form of wage labor. And, under later industrial production, the means of production became increasingly held in the hands of the capitalist class. Moreover, as machinery became increasingly automatic, it further alienated the worker from both their own labor process and the products of this labor. As Marx writes in his "Fragment on Machines," "The science which compels the inanimate limbs of the machinery, by their construction, to act purposefully, as an automaton, does not exist in the worker's consciousness, but rather acts upon him through the machine as an alien power, as the power of the machine itself."[52] In this way, labor power is "alienated" from the laborers who supply it. They no longer have a direct stake in the product or processes of their work; they trade their labor for wages, and what they make is directly alienated from them while their working time is increasingly controlled by machines.

Marxist thinking has made much of this theory of alienation. For example, autonomist Marxism stressed the importance of alienation in understanding social labor and vectors of capitalist exploitation outside the workplace: as capital's reach extends beyond the factory, the target for capital power turns to the entire social edifice—the soul included. Under post-Fordist capitalism, Franco "Bifo" Berardi writes, "No desire, no vitality seems to exist anymore outside the economic enterprise, outside productive labour and business."[53] Alienation in this case is not just the separation of production from the working class, but the alienation of affective production of nearly all life toward capitalist labor and

---

52 Marx, *Grundrisse*, 693.
53 Franco "Bifo" Berardi, *The Soul at Work: From Alienation to Autonomy* (Los Angeles: Semiotext(e), 2009), 96.

consumption. In essence, we become alienated not just in the process of production, but also in our consumption of what advertising sells us and our desire for the life that capitalism presents to us as the image of success.

To resist capitalism solely on account of its alienating power is a mode of resistance that appears beleaguered from the outset, promising only to return us to an unalienated experience of capitalism, rather than an outside—the freedom to be ourselves, but only by buying the proper experiences. In many ways, those who still try to ground resistance to capitalism in opposition to alienation interpret alienation directly through an analysis of the commodity form, hoping for some meaningful and artisanal return to direct production. According to a more traditional Marxism (as well as a whole host of fellow travelers who desire to return to local production), in order to reclaim labor as a fruitful and human-enriching form of social activity, the working class would simply have to repossess the output of production.[54]

It is imperative that alienation is not understood as a unitary phenomenon. "Alienation" as a category has been taken to describe a whole host of losses, from the sometimes-well-paid alienation of professional software engineers, to what scholar of African American literature and history Saidiya Hartman refers to as "natal alienation" caused by the enslavement of Black people.[55] Just how much alienation takes away varies; though it tends to take significantly more, including life, from those who are most marginalized. Moreover, attempts to overcome alienation within capitalism have a way of doing so only for the already well-off, often white men who possess significant capital. For the corporate elite, overcoming alienation can mean significant flexibility, while for the wage laborer, it oftentimes means working in unpredictable fits and

---

[54] See for example Moishe Postone's critique of "traditional" Marxism's focus on capitalism as largely a problem of distribution. *Time, Labor, and Social Domination*.

[55] Saidiya Hartman, *Scenes of Subjection: Terror, Slavery, and Self-Making in Nineteenth-Century America* (Oxford: Oxford University Press, 1997).

starts according to the demands of management. In many ways, the fight against alienation has been appropriated by capitalism, as its superintendents give workers more flexibility and control over their labor, facilely claiming to ease the pressures of schedules by increasing precarity rather than proffering meaningful autonomy.[56] In short, capitalism's reply to workers' complaints of alienation has been to give them the freedom to drive an Uber on their own "unalienated" time. Thus, rather than resist alienation tout court, it is necessary to resist *specific forms* of alienation, with an attention to how, and whom, they take from, and for what outcome. In short, the aim is to neither fetishize alienation nor the imagined sovereignty of a life that could escape all alienation.

In this spirit, readings like that of Moishe Postone, which place at the forefront of their analysis Marx's attempts to establish the abstract character of capitalist society as the source of its particular form of domination, appear more productive and insightful.[57] For Postone, capitalism is characterized by an interdependent set of social relationships that make labor a "structural imperative," a necessary precondition for survival: "No one consumes what one produces, but one's own labor or labor products, nevertheless, function as the necessary means of obtaining the products of others."[58] Under capitalism, our objective social relations demand the necessity of human labor for survival—a condition that is irreducible to the exploitation of laborers by a ruling class. Thus, the reappropriation of one's alienated labor can only result in a superficially improved relationship to work, not the abolition of capitalism's necessity for exploiting labor.[59] In short, as long as capital

---

56 Luc Boltanski and Eve Chiapello, *The New Spirit of Capitalism,* trans. Gregory Elliot (London and New York: Verso, 2005).
57 Moishe Postone, *Time, Labor, and Social Domination: A Reinterpretation of Marx's Critical Theory* (Cambridge, UK: Cambridge University Press, 1993).
58 Ibid., 150.
59 Christian Fuchs and Sebastian Sevignani differentiate the English term work from labor arguing that work is a larger anthropological category of activity that transforms material into useful objects, whereas labor is a form of work specific to capitalist value production. "What Is Digital Labour? What Is Digital

demands the accumulation of wealth among a few, it will produce new compulsions to work and new forms of exploitative alienation that will inevitably fall hardest upon the most marginalized.

Thus, even were we to somehow imagine the replacement of all automatic and algorithmic systems for the production of goods and knowledge with humans who fully understand the production process, this would do little to remove the exploitative conditions of this work and the objectified forces that translate racism, sexism and imperialism into the abstract demands of capital.

To abolish capitalism, what is needed is, rather, a full embrace of the forces of alienation, especially as they are created through machines that automate and extend humanity's ability to produce and think. In short, freedom from the necessity of labor requires a decoupling of machine's productive and alienating force from capitalist accumulation and valuation. The problem humanity faces is not that the products of physical and intellectual labor are wrested away from the factory worker or the scientist, but that their production is aligned with the necessities of a market founded on capital accumulation by the few. Rather than attempt to construct an impossible, unalienated form of capitalism, we must free alienation from capitalism. Again, this does not mean that technology is itself a panacea—for its infernal reproduction promises nothing—but rather that alienation, our very being outside of ourselves in the world, is what must be dissociated from capitalism and repurposed.

In a now-famous segment from his "Fragment on Machines," Marx describes how automation leads to further alienation as not only products but also knowledge itself and the intellectual capacities of laborers are concretized into machines. Marx states, "In machinery, knowledge appears as alien, external to [the laborer] . . . and living labour [as] subsumed under self-activating objectified

---

Work? What's Their Difference? And Why Do These Questions Matter for Understanding Social Media?," *tripleC: Communication, Capitalism and Critique* 11, No. 2 (2013): 237–293.

labour. The worker appears as superfluous to the extent that his action is not determined by [capital's] requirements."[60] Automation renders laborers superfluous as machines become increasingly responsible for production. However, because the abstract dimension of human labor remains the basis of capitalist exchange value and consequently the form of wealth peculiar to capitalist society, capitalism continues to demand the expenditure of human labor in increasingly pointless, and thus even further alienated, work.[61]

Under the privatization and enclosure of the general intellect, the social becomes reformed, erected by the private abstractions of machine learning algorithms and emerging statistical models, wherein the objects that produce the social world—and thus us—are wholly lodged within the capitalist edifice and its alliance with a long history of expropriation and exploitation. Under such a regime, the alienating power of these abstractions can only be in the service of capital accumulation. We work to survive, and the fruits of our labor are given over to capital.

While "alienation" names servitude to various forms of waged and unwaged labor today, its abolition as a category is neither possible nor a guarantee of an end to capitalist exploitation. Instead of combatting alienation through a desire for some external-to-capitalism natural world where there exist conditions for, in Marx's terms, "the full and free development of the individual," we must instead liberate alienation from exploitation. Abstractions—whether the thermodynamic abstraction of productive force into the machine or the abstraction of intellect into computers—are not, by default, oppressive. The collapsing of abstraction into an always-negative alienation effaces the key epistemic conditions that facilitate the possibilities of radical political change.

---

60 Marx, *Grundrisse*, 695.
61 Postone, *Time, Labor, and Social Domination*, 196–197; David Graeber, *Bullshit Jobs: A Theory* (New York: Simon & Schuster, 2018).

Revolutionary Mathematics

Ultimately, the metaphysical, intellectual and political work of a revolutionary mathematics calls for a reconsideration of the relationship between alienation and liberation. This does not mean that these technologies are necessarily liberatory, nor that more technology offers any promise, but rather that the underlying objectifications of the entire trajectory of technology must be contested. The task, then, is to repurpose both the metaphysical ground and the actual capacities of these alienating forms of technology. Such work must seek not to end all production or the alienating force of abstraction, but to engage it oppositionally—to dislocate it from its seeming stability as given or monolithic. In short, we must imagine and construct new mysteries and new modalities of exchange that can enable computation and calculation outside and beyond capitalism. Accordingly, to engage in a revolutionary mathematics is to disavow any belief that we can get under our abstractions, to see "what is really there" and regulate away their problems. Instead we must reckon with this world of abstraction and alienation.[62]

This requires us to create and build the world from different mysteries and different exchangabilities: ones that deny the power and reality of imperialism, and even the most basic form of capitalist exchange, by which one trades their work for the capacity to continue living. We require different means of counting and valuing. A revolutionary mathematics aims not to recoup or destroy those old forms through critique, but to create different abstractions—even new "natures" and the valorization of natures outside the Enlightenment notion of individual sovereignty—and with them new alienations arising from new, mysterious metaphysics of exchange that offer, just like the commodity and the machine, to

---

[62] The xenofeminist collective Laboria Cuboniks argues: "The radical opportunities afforded by developing (and alienating) forms of technological mediation should no longer be put to use in the exclusive interests of capital, which, by design, only benefits the few." *The Xenofeminist Manifesto: A Politics for Alienation* (London and New York: Verso, 2018), 35.

think for us, both economically and computationally (which today amounts to the same thing). To do this does not mean to simply disregard old forms, for it is only by tracing their histories and implications—as the present text has attempted to do—that it may be possible to build new mysteries that are not merely repetitions of the present.

The Bayesian discovery that probabilistic knowledge depends on markets and exchange appears initially to naturalize capitalism. In the final analysis, however, the result is the exact opposite: Bayesian statistics denaturalizes nature and calls on any who value knowledge, or even the possibility of knowing, to reconceive the very notion of exchange and its metaphysical ground. A revolutionary mathematics must oppose the naturalism of the unalienated, epistemic subject, and the fantasy of the liberal subject who can efficiently operate the material and economic world within which they find themselves; in short, such a possibility opposes both the naive chauvinism of Ronald Fisher and the attempts to naturalize economism of Leonard Savage and Bruno de Finetti.

A new revolutionary object, or objectification, that would break the stranglehold of the commodity on our thinking must also abandon the fetishism of the natural, the unalienated and the dereified. At the same time, it must forswear the possibility of any guaranteed future. We cannot return to some before or some future where we finally grasp the totality of our situation. A revolutionary mathematics must commit to the alien and the mediated, but this declaration means something very specific within the context of mathematics and machine learning: namely, that we must reject the Fisherian residue that offers us a fetishistic version of knowledge production by claiming that knowledge belongs to the individual and that it is immune from the material and economic conditions of its production. In sum, this means we must collectively reimagine the production and meaning of scientific and technical knowledge.

The Bayesian revolution has taken recourse to the exchange of contracts in order to provide a foundation for probabilistic knowledge, but in making knowledge second to exchange, it fundamentally denaturalizes and subjectivizes knowledge, abandoning "objective" theories of probability. This revolution does away with claims that knowledge directly represents the physical world as it is, replacing it with a representation of the world *as it is profitable*—in short, knowledge now finds its future in the subjective and the social. In doing so, it requires knowledge to admit its fundamentally material and economic ground and permits a demonstration of the fact that the very systems of exchange from which it arises can no longer support its function. For scientific knowledge to have a future beyond its current capitalist crisis, we must accept knowledge's abstract and alien power.

Additionally, the work of a revolutionary mathematics and its attempts to create new objectifications is a politics that leaves the subject alone, for it has little need for the idea of a fixed and solid subject. This politics operates against the presumption that solidarities can be either forcefully manufactured from above—for instance, via the enforced class dynamics of Leninism, which tries to tell the revolutionary subject how to be—or organically produced from below—via some magical event that would invigorate an organic, collective solidarity. This politics detaches from the desire to make subjects and instead focuses on the metaphysical technics of producing objectifications and, hence, on what is objectively the case. For example, those who think of big data in terms of privacy are half right: big data's desire to make our subjective relationships computable is one that should be rejected. But to stop there is to lose a losing battle. It is to play with a coin that is wholly biased toward capital. Instead, we must fundamentally reconceptualize what we are computing and why. While such calls for privacy are clearly important, to simply show that we are not being valued equally or that we are being exploited will never be enough to change our condition.

It would be ideal to be able to offer examples, to point to specific ideas or individuals who are currently advancing this work. But such a revolution in thought can only be fully ascertained and traced after the fact. What may appear revolutionary today could, in the hindsight of tomorrow, appear fully reactionary. Yesterday's provocative and revolutionary attempts to stand outside of digital capitalism appear today as withdrawal and retreat. Still, the act of tracing the grounds upon which our current epistemology and metaphysics of exchange stand—while refusing to think that we will ever discover some unalienated truth of how things "really are"—can reveal unknown contradictions and new demands. It is out of these demands for a different world that the work of revolutionary mathematics progresses. It is far from clear what these new or different metaphysics could be, but the task ahead is to seek them out.

If knowledge production and its metaphysical ground are founded upon the historical and material conditions of exchange, to politically engage these processes requires that we intervene in the process of exchange itself. But let us be clear: this does not mean that politics should be aimed exclusively at capitalism or class conflict (in the traditional Marxist sense). To the contrary, every inequality we face today—sexism, xenophobia, racism, transphobia, ableism—constitutes a mobile and multifaceted system that defines the varied axes of exchange, oppression and injustice.[63] All of these socially constructed axes of oppression allow computation and exchange to function—providing various forms of exploited labor, such as the outsourcing of the psychic trauma of content moderation—and replicate themselves in the computed outcomes of various technologies of machine learning and statistics.[64] Discrimination, oppression and injustice are not

---

63   See Ernesto Laclau and Chantal Mouffe, *Hegemony and Socialist Strategy* (London: New Left Books, 1985).

64   Adrien Chen, "The Laborers Who Keep Dick Pics and Beheadings Out of Your Facebook Feed," *Wired*, October 23, 2014; Mary Gray and Siddharth Suri, *Ghost Work: How to Stop Silicon Valley from Building a New Global Underclass* (New York: Eamon Dolan Books, 2019); Sarah Roberts, *Behind the Screen: Content Moderation in the Shadows of Social Media* (New Haven, CT: Yale University Press, 2019).

reducible to exchange but rather are operationalized and exploited through exchange, replicating their violence in the knowledges that computation produces. The task is to directly confront these injustices wherever they may arise—but without claims to a return to an unalienated or natural world—and in doing so to trace the various metaphysics, forms of knowledge production and material economies that make them thinkable and computable today and, perhaps, make a future beyond them imaginable.

The work of the revolutionary mathematician is, in the end, less mathematical than it is metaphysical. It is not a simple decision or discovery that a sovereign liberal subject could conjure from nowhere. While we are far from knowing exactly where this work leads or exactly what form it must take, the task is to strategically create new mysteries and resist injustices as constitutive parts of this amalgamation of capitalist computation. Like those early founders of calculus who created a mathematics that worked without fully understanding its foundations, the work ahead is to seek out new and different equivalences, values and computabilities, all the while understanding that each of these is fundamentally social, abstract, alienatory and unnatural. The path backward toward some natural redemption closed long ago, if it was ever open. A revolutionary mathematics requires that we create new objectifications that call upon us to build a different future, whether we believe in them or not. In the face of a global capitalist system committed to accumulation at the cost of exploitation and the destruction of the ecosystem, nothing could be more difficult. But at the same time, nothing could be more necessary.

Conclusion: Toward a Revolutionary Mathematics

Right-wing billionaire Robert Mercer's career is intimately tied to everything that is at stake in the Bayesian revolution. In 1972, after completing his PhD in computer science, he joined IBM's research division. There, he helped make major advances in speech recognition and machine translation. Few on Mercer's team had much of a linguistics background, but their efforts to apply probabilistic models to speech nevertheless succeeded, and now these form the basis of many of our contemporary algorithms for working with human language. Notably, Mercer's team developed a Bayesian-informed algorithm for determining the probability that a given string of words will be observed—even if it has never been observed before.[1] For this work, Mercer was awarded a lifetime achievement award from the Association of Computational Linguistics in 2014.[2]

In 1993, Mercer was lured away from this research to join Renaissance Technologies, a hedge fund that employed statistical and machine learning techniques to direct their investments.

---

[1] Peter F. Brown, Peter V. deSouza, Robert L. Mercer et al., "Class-based N-gram Models of Natural Language," *Computational Linguistics* 18, No. 4 (1992): 467–479.

[2] "Robert L. Mercer Receives the 2014 ACL Lifetime Achievement Award," Association for Computational Linguistics official website, October 15, 2014.

Mercer brought his Bayesian methods from IBM with him, developing investing algorithms that helped the fund generate returns averaging 39 percent between 1989 and 2006.[3] In 2009, he became co-CEO of the firm, a position he held until his retirement in 2017. By using similar probabilistic techniques to those he developed at IBM, Mercer was able to amass billions of dollars in private wealth, both for his clients at Renaissance Technologies and for himself.

But Robert Mercer's story does not end with this capital accumulation. Since the early 2000s, Mercer—both individually and through his family's foundation—has donated tens of millions of dollars to conservative causes.[4] He has funded the right-wing Breitbart News Network, groups opposing the so-called "Ground Zero Mosque" in lower Manhattan, and climate change denial think tanks like the oddly named Berkeley Earth group.[5] He also provided the initial funds to start the now-infamous Cambridge Analytica consulting firm, which obtained data from Facebook in order to target political ads supporting Donald Trump's 2016 election campaign.[6]

We witness, here in miniature, how capitalist privatization is able to turn the promise of science and the development of the general intellect against itself. Mercer has pioneered some of the very methods that have driven late twentieth- and early twenty-first-century science and knowledge production. Yet the fruits of those methods were extracted in the form of private wealth, and then turned against science in favor of a populist and racist nationalism. Again and again, the motives of capital accumulation and monopolization reward the simultaneous privatization and privation of general knowledge.

---

3 Zachary Mider, "What Kind of Man Spends Millions to Elect Ted Cruz?," *Bloomberg*, January 20, 2016.

4 "Mercer Family Foundation," Conservative Transparency official website.

5 Robert Pogrebin and Somini Sengupta, "A Science Denier at the Natural History Museum? Scientists Rebel," *New York Times*, January 25, 2018.

6 Carole Cadwalladar and Emma Graham-Harrison, "Revealed: 50 Million Facebook Profiles Harvested for Cambridge Analytica in Major Data Breach," *Guardian*, March 17, 2018.

In the final analysis, our current array of deep epistemological crises—from the replication crisis in the sciences, to concerns about fake news, to the existence of defeat devices, to filter bubbles that insulate individuals from contrasting views online—are at their heart crises of capitalism. As neoliberal capitalism ceases to focus on the management of production and instead turns to the management and control of knowledge, it risks a whole new series of crises. As Paulo Virno writes, "The models of social knowledge do not turn varied labouring activities into equivalents; rather, they present themselves as 'immediately productive force.' They are not units of measure; they constitute the immeasurable presupposition of heterogeneous effective possibilities."[7] But in making these knowledges directly productive, capitalist knowledge production sweeps the ground out from under its own feet. While capitalists believe its mysteries—their unequal equalities—are directly knowable, they privatize and deprive these knowledges of their collective ground.

On a metaphysical level, statistics and economy share the same goal: to relate particulars to universals. Statistics aims to deduce general laws from individual pieces of data: it proceeds, for example, from individual demographic data to larger social trends. Economic systems do the inverse: they produce individual acts of exchange from general principles, increasing production here and laying off workers there as markets fluctuate. But machine learning—especially given that it advances, in the words of former *Wired* editor Chris Anderson, "without coherent models, unified theories, or really any mechanistic explanation at all"—seeks only to relate the particular to the particular, or what is the same: to make every particular universal, sending an advertisement or service to a user at precisely the moment they require it.[8]

In this supposed theory-less world, science is only able to relate data to its most proximate step: what should be done next. As

---

7 Paolo Virno, "General Intellect," *Historical Materialism* 15, No. 3 (2007): 6.
8 Chris Anderson, "The End of Theory: The Data Deluge Makes the Scientific Method Obsolete," *Wired*, June 23, 2008, 16–17.

Leonard Savage argued, the key question in science ceases to be what to say, and instead becomes what to do. The only possible, and desirable, knowledge becomes that of how to act in the immediate moment, while larger questions of political economy, and the possibility that things could be different from what they are, are ignored. This, then, is the mathematics of capitalist orthodoxy. Here we see how the more machine learning succeeds in its predictive power, the more it fails in its ability to efficiently know the world or even to distribute capital, for it sweeps away the ground upon which capitalism claims to be able to manage production. The science behind machine learning creates massive incentives not to solve problems or develop the general intellect, but rather to game the system and enclose knowledge.

This failure of knowledge production is, then, a failure to account for the political economy that lies both in the metaphysical core and the practical uses of machine learning and statistics. Knowledge produced from the exchange of contracts ends in the exchange of contracts, with the only meaningful goal being the avoidance of the Dutch book. One is left, as Marx says, deciphering the very hieroglyphics the market has left there. But this subjectivization of knowledge does not make statistics useless. Rather, these methods offer to produce abstractions that are collectively productive, providing for the world rather than exploiting it. But to do so, they must be freed from the necessities of capital accumulation and its constant drive to profit from racism, sexism and imperialism.

As machine learning and statistics cut away their own roots, disavowing the Fisherian belief in the heroic individual scientist, they threaten to shake the mysteries of Enlightenment scientific knowledge production to their core. Where there were once solid equivalences between labor and knowledge, these directly productive probabilities once again set everything in motion. As Virno writes,

> The principle of equivalence used to be the foundation of the most rigid hierarchies and ferocious inequalities, yet it ensured

a sort of visibility for the social nexus as well as a simulacrum of universality. This meant that, albeit in an ideological and contradictory manner, the prospect of unconstrained mutual recognition, the ideal of egalitarian communication and sundry "theories of justice" all clung to it.⁹

Virno continues by arguing that the collapse of this principle of equivalence is now the cause for cynicism: "The cynic recognises the primary role of certain epistemic models, as well as the absence of real equivalences. He sets aside any aspiration to transparent and dialogical communication. From the outset, he relinquishes the search for an intersubjective foundation to his praxis or a shared criterion of moral judgement."¹⁰

This is precisely the dual danger and opportunity of machine learning. In making knowledge directly productive, machine learning undermines the foundation of this principle of equivalence. Virno's definition of the cynic aptly describes Savage's position (and Alfred Sohn-Rethel's critical apperception of this fact): statistics only works in the context of exchange, and the more we pursue real equivalents—such as the objectivity of the frequentists—the faster they evaporate. With no stable ground, these age-old metaphysical concepts' very difference is computed anew at every moment.

The rationality that these methods and technologies embody demonstrates a codependence that is unable to separate the objective and the subjective, because each collapses into the other. Given that the very ideal of objectivity can only be sustained by one's subjective belief in that ideal—like the ideal coin of frequentism and capitalism—the whole system of knowledge production comes to insist on the centrality of exchange and, with it, social relations. In this way, the Bayesian revolution has made the fluid and mobile economics of a neoliberal hyper-capitalism central to the

---

9 Virno, "General Intellect," 6–7.
10 Virno, "General Intellect," 7.

production of objective knowledge. The social relations of exchange that appear to many commenters as a corruptive influence on statistics and the production of knowledge—the contamination of "pure science" by the soiled touch of human influence—lie at its metaphysical heart.

We are witnessing, then, a new era, a new version of what Marx means when he writes that "all that is solid melts into air." Or as he says in the "Fragment":

> Everything that has a fixed form, such as the product etc., appears as merely a moment, a vanishing moment, in this movement. The direct production process itself here appears only as a moment. The conditions and objectifications of the process are themselves equally moments of it, and its only subjects are the individuals, but individuals in mutual relationships, which they equally reproduce and produce anew.[11]

It is for this reason we must, in a way, side with Leonard Savage and Bruno de Finetti, and even to an extant with Jerzy Neyman and Egon Pearson, but in order to take their discoveries further: while they appear as quintessential capitalists, their insights allow us to reject Ronald Fisher's fetishism of knowledge as a product that can be owned by the lone scientist toiling in the lab. Through their emphasis on exchange, we are able to see the economically driven processes that create these knowledges and the modes of production they support. These latter statistics reveal the extent to which knowledge depends on political economy and the necessity of working through crises of economy if we are to address crises of knowledge production.

This turn to process offers no promise that we will be free of objectification or its alienating powers, or from sadistic forms of concrete domination. Rather, we witness these new mobile and

---

11 Karl Marx, *Grundrisse: Foundations of the Critique of Political Economy* (London: Penguin, 2005, repr.), 712.

temporary assemblages of objectification pulling us in all directions, as they did Mercer, who produced an anti-rationality that is the exact opposite of the rationality he had commodified only yesterday. But, in the face of these destabilizations, we are able to steal a glimpse, as Marx shows us, at the process rather than simply the product.[12] It is through this recognition of process, in all its social implications, that a new objectification is possible.

At the same time, we must recognize that the border that separates the particular and the universal is at stake. These technologies, and thus objectification writ large, are becoming the very forces that shape and define the universal—a universal that excludes particular subjects and individuals. The technologies and methods of machine learning, data analysis and statistics offer us a means by which we could wholly reconfigure the relation between individual and universal in modes that far outpace the violence and destruction of capitalism. To do so requires that we think through and reconfigure the very forms of objectification through which they work and disavow the fantasy of complete understanding or a return to some prior more natural state. Likewise, they require that we disavow the promise of any future guarantee and the fetishism of the new for its own sake—both teleological principles that all too often repeat the dream of a male Eurocentric universalism.

As neoliberal capitalism seeks out more and more local contexts from which to extract value and knowledge, it invests in the very dissimulation that undermines its claims to efficiency and objectivity. Machine learning and statistics are now in the process of creating a new set of algorithmic objects that are unseating the commodity's economic centrality, as value appears to derive directly from automatic computation. Even as they build "real abstractions" on top of the current metaphysics of exchange, these algorithmic objects demonstrate themselves to be anathema to their enclosure

---

12 Georg Lukács, *History and Class Consciousness: Studies in Marxist Dialectics*, trans. Rodney Livingstone (Cambridge, MA: MIT Press, 1967), 259.

and privatization—for their enclosure under conditions of capitalism guarantees that they will continue to become technologies of deception rather than production, unlikely to be constrained by calls for regulation. Even if, at the moment, another possibility sounds only as a faint whisper, these technologies—in accounting for our affairs—call out for collectivization: for shared ownership of both the means and metaphysics of production.

To seize the means of production is both a political and metaphysical task. Production, whether industrial or statistical, produces both objects and objectification. If we seize only the means of producing the former, we will, in the end, only reproduce the logic we hope to escape. Although its exact outline is still far from clear, this requires that we trace and understand the current metaphysics of capitalism—not in order to dereify them, but rather to understand how they can be replaced in order to valorize new and different forms of knowledge and value.

To follow this path is to become revolutionary mathematicians—to work on the level of metaphysics, creating new and different equalities based on new and different mysteries. This aim cannot be confused with some futurism that would forsake the past; indeed, that would be impossible, for the tradition of all dead generations weighs like a nightmare on us, and we carry its weight along with us. To ignore this weight would simply be to allow it to repeat itself. We must instead attend to history—its forms, its contradictions and those people and things about which it calculates. These chapters have sought to show that it is possible to transform and to work on the mathematics and metaphysics that constantly calculate this weight and its value, altering with it the very divide between the subjective and objective. For it is only then that we can objectively turn our world of algorithmically mediated exploitation—of environmental destruction, abuse of workers, racism, sexism, heterosexism, ableism, xenophobia—into one that is more just and more equal, in a sense far beyond the limited equality of capitalist commodity exchange.

Index

A
abc conjecture, 62–6
abstract domination, 13–14, 94, 201
abstractions. *See also* real abstraction
  algorithmic abstractions, 185–8
  freedom in, 178–85
  as increasingly mobile and modulatory, 177
  liberatory potential of, 205n50
  new forms of, 202
  as not necessarily a force for exploitation, 205
  power of commodity as, 180
  productive tensions of with location, 181
  statistical models as providing mathematical abstractions of the world, 177
  as tending to generate authority, 180
activation function, in machine learning, 47
Actor Network Theory, 83n9
Adorno, Theodor, 180
AGI (artificial general intelligence), 43n11
AI (artificial intelligence)
  AI winter, 4, 53
  as not living up to postwar promise, 52–3
  as still only able to solve problems that are given to it, 201
algorithmic bias, 44
algorithmic instructions, as symbol manipulation, 78
algorithmic knowledge
  consequence of, 186
  limits of, 38
  need for revolutionary approach to, 33
  presumed veracity of, 6
  preventing worst abuses of, 61
algorithmic logic
  advances in, 7
  inequities of, 14
  presumptions of, 6
  of Walmart and Northpointe, 44
algorithmic models, 22, 41, 78
algorithmic power, 81
algorithmic society, 8, 29–30, 34
algorithmic systems
  attempts to slow functioning of, 59
  as controlling much of contemporary life, 10, 14
  filter bubble of, 16
  as functioning in complex ecologies, 61
  inaccuracy of, 200
  as relative to data put into it, 41
  as supposed to achieve real-time abstractions, 196
  as thinking for us, 101
algorithmic trading, 4, 191
algorithms
  as allowing conversion of extractable data into interchangeable bits, 182
  classification algorithms, 51
  defined, 77–8
  as effectively laundering nonobjective forms of violence and bias, 174
  foundation of, 13
  inaccessibility of, 59–60
  as more efficient at creating new realities than in representing a world, 7
  and objectification, 21, 77–98
  as opaque, 79
  predicting likelihood of recidivism, 44
  regulation of, 60, 61
  role of in mediating society and economics "behind our backs," 20
  self-correcting design of, 38
  training of, 42
  as turning data from the world into knowledge about that world, 30
  use of for obfuscation, 61

Index

alienation, 206–11
Alpaydın, Ethem, 49
alternative hypotheses, 119, 126, 130, 138
Althusser, Louis, 2n1, 80n4
Amaro, Ramon, 92–3, 182n9
American Statistical Agency (ASA), on p-value, 118
The Analyst (Berkeley), 32, 67, 70
Anderson, Chris, 56, 57, 58, 75, 219
Antipater of Thessalonica, 99, 113
Appel, Kenneth, 62, 63, 66
Arbuthnot, John, 111, 115, 163
artificial general intelligence (AGI), 43n11
artificial intelligence (AI)
  AI winter, 4, 53
  as not living up to postwar promise, 52–3
  as still only able to solve problems that are given to it, 201
artificial neural networks (ANNs), 45–52, 53, 55, 186
The Art of Computer Programming (Knuth), 77
automated abstractions, and alienation, 177–216
automation
  in algorithmic trading, 4
  of knowledge, 37–58
  role of, 28
autonomism, 25

B

backpropagation, 48, 49, 55
Bailey, Arthur, 164
Baucom, Ian, 181n7
Bauman, Zygmut, 189n15
Bayes, Thomas, 67, 143, 163, 164
Bayesian approaches/statistics
  as admitting to theistic understanding of probability, 162
  as claiming to avoid reference class problem, 40
  as denaturalizing nature, 213
  as flipping frequentism's metaphysical perspective on its head, 158
  as fundamentally changing stakes of statistical analyses, 160
  how they function, 94
  methods of as ascendant, 12
  naive Bayes classifier, 55, 153–7, 161–2
  as potentially leading to maximization of profit, 173
  probability as measure of subjective belief, 148
  and problem of frequentism, 143–57
  as requiring use of probability distribution, 150
  revolutionary implications of, 183
  rise of, 144, 149, 160
  shift from frequentist to, 30–1
  subjectivism of, 155, 156, 160
  value of, 148–51
Bayesian inference, 152
Bayesian metaphysics, 158–76
Bayesian revolution, 12n21, 34, 156, 162, 174, 199, 205, 213–14, 217, 221–2
Bayes's theorem, 143, 149, 151–2
Bennett, Craig, 109, 110
Berardi, Franco ("Bifo"), 17, 31, 207
Berkeley, George, 32, 67–8, 69–71, 72, 74, 75
Berryman, John, 35, 43
Bible, on lots and chance, 112n20
big data
  benefits of according to advocates, 6
  use of to present appearance of following rules while wholly disregarding them, 6
Bitcoin, 94–8
black boxes, 59, 61, 186
Black Marxism (Robinson), 91
blockchain, 94–8
Boolean Pythagorean triples problem, 62–3
Bowker, Geoffrey, 57
Bristol, Muriel, 115–16, 120, 125
Browne, Simone, 14, 29
Bruno, Giordano, 164n12
Bryant, Levi, 164n12

C

calculus
  Berkeley's criticism of, 32, 67, 70, 71, 72
  rationale for teaching of, 69–70
Canadian Tire, use of data to predict customer behaviors, 44
Cantor, Geoffrey, 72n22
Capital (Marx), 19, 80, 84
capitalism
  abolishment of, 210
  according to Postone, 13, 209
  algorithmic knowledge production as unable to keep pace with motives and drives of, 4
  calls for retrenchment of an increasingly privatized capitalism, 3
  capitalist calculation problem, 195–201
  capitalist knowledge production, 219
  cognitive capitalism, 26
  as constructed atop abstractions that separate concepts from context, 179
  crisis, 33, 197, 219
  digital capitalism, 21, 26, 151, 189, 194, 198
  as founded on metaphysical process, 19–20
  frictionless capitalism, 183
  importance of statistics to understanding contemporary capitalism, 8
  industrial capitalism, 26
  informational capitalism, 12, 34, 149, 154, 156, 161, 171
  information as new frontier of, 190
  machine learning as function in way analogous to, 58
  machine learning systems as working to solidify, 58
  need to free alienation from, 210
  as objectifying, 33
  as operating as algorithm, 84–5
  platform capitalism, 26, 192–3
  relation of statistics to, 12–13
  as rendering labor increasingly superfluous, 96–7
  rise of algorithmic capitalism, 7
  short-term profits of as blocking any real response to climate change, 40
  split with scientific production, 174
  statistics' metaphysical role in, 10
  as subsuming and transmuting all social relations into modes of production, 198–9

Revolutionary Mathematics 227

as supporting injustice and exploitation, 75
surveillance capitalism, 26
understanding nature of contemporary capitalism, 20
Carceral Capitalism (Wang), 91
Cauchy, Augustin-Louis, 68n13
Centers for Disease Control and Prevention, on doctor visits for flu, 41
Chichilinisky, Graciela, 170
Chun, Wendy, 9n18, 30n55, 57–8
class exploitation, as form of concrete domination, 92
classification algorithms, 51
climate change, capitalism's privileging of short-term profits blocking any real response to, 40
cognitive capitalism, 26
colonialism
 consequences of, 190
 as constructed atop abstractions that separate concepts from context, 179
commodities, and objectification, 84–8
commodity exchange, calculation as taking on form of in Bayesian analysis, 164
The Communist Manifesto (Marx), 15
COMPAS (Correctional Offender Management Profiling for Alternative Sanctions), 44
computation, role of in determining current objective social reality under digital capitalism, 21
computers, can they do math? 59–76
concrete violence, 94
correlationism, 75
Cox, Sean, 4–5
cryptocurrencies, 96
cypherpolitics, 95

D
Davies, William, 22
Dean, Jeff, 185
Dean, Jodi, 17
deep learning, 185
defeat device, 4–5, 60, 64, 194, 195, 219
De Finetti, Bruno, 164–5, 168, 169, 170, 171, 189, 213, 222
Deleuze, Gilles, 189n15
Demeter, 113
Dependency Road (Smythe), 88n16
dereification, 17, 22, 60, 69, 70, 90
Derrida, Jacques, 70n17
The Design of Experiments (Fisher), 115
deviancy, measurement and management of, 29
digital capitalism, 21, 26, 151, 189, 194, 198
digital systems, as providing ideological form for contemporary neoliberalism, 30n55
Dillman, Linda, 37, 38
divine connection, 163
Dutch book argument, 164–76, 183, 220
Dyer-Witheford, Nick, 28, 194

E
economy. *See also* political economy
 digital systems as fundamentally altering function of, 23
 goal of, 219
 knowledge economy, 199

The Eighteenth Brumaire (Marx), 8
Elia, Ramón de, 102
enclosed intellect, 195
enclosure, of general intellect, 188–95, 201, 211
Endnotes, 18
Enlightenment
 abstraction of, 182
 as constructed atop abstractions that separate concepts from context, 179
 idealism of, 183
 as inspiring varieties of communism, 15
 models and theories of causality in, 56
 notion of individual sovereignty in, 212
 rise of concept of race during, 58
epistemic authority, 73–6
An Essay towards Solving a Problem in the Doctrine of Chances (Bayes), 162
Eubanks, Virginia, 29
eugenics, 121
exchange
 belief, experience, and subjectivity as conditioned by constraints of, 204
 early deistic forms of abstraction as bearing some relation to, 179n3
 notions of justice and freedom as founded on, 181n7
 tethering of knowledge to, 171–2
 victory of neoliberalism as means to organize all life and subjectivity on grounds of, 172
extrapolation, limitations of, 45

F
Facebook
 News Feed algorithm, 40
 as providing secret access to user data to large corporate partners, 189
 as taxing exchanges, 193
face recognition, 182n9
fake news, 219
false negatives (errors of the second kind), 111, 124, 126, 128
false positives (errors of the first kind), 107, 111, 124, 126, 127, 128, 151, 152
falsifiability, 123
Ferguson, Scott, 205n50
fetishism, 80n4, 89n19, 138, 206, 209, 213, 222, 223
filter bubbles, 16, 219
First Vatican Council, 71
Fisher, Ronald, 9, 115–18, 119–20, 121–3, 124, 128, 129–30, 135–6, 137–8, 139, 144, 150, 164, 172, 173, 174, 175, 193, 194, 213, 220, 222
Fisher-Neyman-Pearson model, 131
fMRI (functional MRI)
 cluster failure problem in, 106–9, 124, 132, 134–5
 structural problems within statistical methodologies of, 109
Foucault, Michel, 28
Foundations of Statistics (Savage), 169, 170
four-color theorem (4CT), 62–5, 66
"Fragment on the Machine" (Marx), 25, 207, 210–11, 222
Frankfurt school, 180

Franklin, Seb, 30n55
free labor, debate about existence and nature of on digital platforms, 24
freethought/freethinkers, 71, 72, 73–4, 180
frequentist statistics/frequentism
  as answering wrong question, 147–8
  Bayesian statistics as flipping metaphysical perspective of, 158
  compared to Bayesian analysis, 183
  emergence of, 104–5
  on probability, 104–5
  problems with, 161
  reference classes as pivotal for, 145–6
  shortcomings of, 144–5, 147–8
  as struggling, 143
Friedman, Milton, 50n22
Fuchs, Christian, 24n43, 209n59
functional MRI (fMRI)
  cluster failure problem in, 106–9, 124, 132, 134–5
  structural problems within statistical methodologies of, 109

G
Gates, Bill, 183
Gelman, Andrew, 57n39, 125, 134, 171
general intellect
  defined, 191
  enclosure of, 188–95, 201, 211
The Genetical Theory of Natural Selection (Fisher), 121
ghosts of departed quantities, 68, 69, 73, 75, 87, 95, 96, 138
Giannandrea, John, 43n12
Gigerenzer, Gerd, 114n23, 172n23
Goertzel, Ben, 43n11
Goldhaber, Maurice, 133
Google
  and advertising world, 56
  attempt to predict flu activity based on search results, 41
  "child" algorithm, 51n23
  image recognition system, 185–8
  PageRank algorithm, 78, 183–4
"Greyball" program (Uber), 5–6
Grundrisse (Marx), 96

H
Haken, Wolfgang, 62, 63, 66
Hartman, Saidiya, 208
Harvey, David, 190, 198
Hayek, Friedrich, 195–7, 200
Hegel, G.W.F., 49, 56
Henry I (king), 82
homophily, 57–8
Honneth, Axel, 83n11
Horkheimer, Max, 180
Husserl, Edmund, 164n12
hybridization, 130–5, 137, 139
hyperparameters, 50
hypothesis testing, 111–12, 113, 115, 117, 118, 124, 134, 151, 155, 156

I
ideal coin, 10–11
image recognition system, 185–8
immaterial labor, 24, 25
imperialism
  capital accumulation's constant drive to profit from, 220
  as form of concrete domination, 92
  machine learning systems as working to solidify, 58
  objectified forces as translating of into abstract demands of capital, 210
  as supporting injustice and exploitation, 75
  as translated into abstract demands of capital, 210
incommensurability, 183
individualism, 172, 173, 174, 193, 220
inductive behavior, 125–6
industrial capitalism, 26
inference revolution, 10, 172n23
infidel mathematician, 67–73
informational capitalism, 12, 34, 149, 154, 156, 161, 171
intellectual property rights, 190n20
interpolation, successes of, 45
inverse probability, 150, 151, 154
inverse problem, 148, 149–50
Ioannidis, John, 135, 139

J
John Deere, 192
Jonah (biblical), 110–14
Jurin, James, 74, 180

K
k-nearest neighbors, 51n25
Knight, Frank, 39
knowledge
  automation of, 37–58
  contradiction of in digital capitalism, 194–5
  foundation of, and Bayesian metaphysics, 158–76
  inference revolution as having revolutionized production of, 10
  machines as producing, 27
  movement from objective to subjective foundations for, 12
  privatization and collectivization of, 194
  as quasi-independent productive force, 198
  tethering of to exchange, 171–2
knowledge economy, defined, 199
knowledge production
  algorithms as changing speed and form of but not ultimate goals of, 61
  Bayesian statistics as fundamentally altering, 156
  capitalist knowledge production, 219
  engagement of, 206
  failure of, 220
  as founded upon conditions of exchange, 215
  as insisting on centrality of exchange, 221
  Robert Mercer's role in, 218
Knuth, Donald, 77–8
Kobe Steel, 200
Kornbluh, Anna, 205n50
Kotsko, Adam, 205n50

L

labor
- capitalism as rendering labor increasingly superfluous, 96–7
- forms of that work on data, information, and knowledge, 24–5
- free labor, 24
- immaterial labor, 24, 25

Laboria Cuboniks, 212
Lacan, Jacques, 18
lady tasting tea experiment, 115–18, 119, 120, 125, 132
Laplace, Pierre-Simon, 117, 141, 143, 149
Laprise, René, 102
Latour, Bruno, 83n9
learning rate, 49, 50
Lenin, Vladimir, 15n27
Lindley, Dennis, 144, 145
local optimization, 201
"A Logical Calculus of the Ideas Immanent in Nervous Activity" (McCulloch and Pitts), 53
Lukács, Georg, 17, 82
Lynch, Michael, 149

M

machine learning
- activation function in, 47
- as appearing to establish precisely what capitalism has always dreamed of, 182
- challenges to, 41
- crisis in, 4
- current uses of, 199
- dual danger and opportunity of, 221
- as functioning in way analogous to capitalism, 58
- goal of, 219
- as mobilizing and liquifying abstractions, 182
- newer methods of, 55
- operationalization of statistics in, 189
- *operation of*, 38–9
- origin of, 52
- preference for correlations over causal explanations of, 181–2
- problems in, 51n24
- relationship of to capitalism, 41n8
- science behind, 220
- and statistical modeling, 41–2
- successes of, 43

Marewski, Julian, 114n23, 172n23
the market, as metaphysical ground, 162–4
Martin, Randy, 97
Marx, Karl, 1, 7–8, 15, 19–20, 25, 27, 67, 73, 80, 81, 83, 84, 85–6, 88, 90, 96–7, 139, 190, 191, 206–7, 209, 210–11, 220, 222, 223
Marxists/Marxism, 3, 15, 18, 19, 22, 193–4, 207, 208
mass intellectuality, 193
mathematical, as level of statistics, 8–9
mathematics. *See also* revolutionary mathematics
- can computers do math? 59–76
- of capitalist orthodoxy, 220
- theoretical mathematics, 62–5

McCulloch, Warren, 53–4
Mercer, Robert, 217–18, 223
meta-analyses, 160n3
metaphysical, as level of statistics, 8, 9
metaphysics
- Bayesian metaphysics, 158–76
- capitalism as founded on metaphysical process, 19–20
- the market as metaphysical ground, 162–4
- metaphysical force of objectification, 58
- metaphysical foundation of statistics, 122
- objectification as metaphysical process, 88
- objective metaphysics, 88–90
- statistics' metaphysical role in capitalism, 10
- work of revolutionary mathematics as metaphysical, 33

Minsky, Marvin, 53, 54
Mochizuki, Shinichi, 64–5
Moten, Fred, 97
Mumford, Lewis, 74n24
mysteries
- of faith, 71–2
- linkage of equality and inequality in, 75
- of reason, 76

N

naive Bayes classifier, 55, 153–7, 161–2
natal alienation, 208
naturality, and ontology, 201–6
negative results, academic movement as organizing for sharing of, 137
Negri, Antonio, 20n40, 194n28
neoliberalism
- digital systems as providing ideological form of, 30n55
- Hayek as founding, 196
- political and social order as increasingly subject to constraints of, 3
- in scientific thinking, 129
- as seeking out more and more local contexts from which to extract value and knowledge, 223
- victory of as means to organize all life and subjectivity on grounds of market exchange, 172

network society, 26
neural networks, 53, 54, 55, 61, 182–3. *See also* artificial neural networks (ANNs)
neurons
- artificial "neurons," 45
- hidden neurons, 46, 47, 50

News Feed (Facebook), 40
Newton, Isaac, 70, 72, 74, 133, 180
Neyman, Jerzy, 102–3, 123–30, 136, 137, 138, 139, 144, 156, 164, 169, 172, 175, 189, 222
Neyman-Pearson model/system, 126, 131, 136, 169
Noble, Safiya, 29
nonlinear dynamics, 48, 50, 202
Northpointe, 44
Norvig, Peter, 155
null hypothesis, 119, 120, 122, 123, 124, 126, 129–30, 144, 147–8

O

objectification
- according to Marx, 19–20
- algorithms and, 21
- algorithms of, 77–98

Bayesian approach as taking process of more seriously than frequentism, 158–9
commodities and, 84–8
creating and exploring new possibilities of, 34
defined, 20
as form of forgetting, 83
God as force of, 164
how subjective knowledge is made to appear objective, 164n12
as means through which concrete domination and violence are given abstract form and then translated back again into the concrete, 94
metaphysical force of, 58
as metaphysical process, 88
mobile and temporary assemblages of, 222–3
mysteries of, 73, 97
as not a blueprint, 91–4
as not necessarily bad or unacceptably reductive, 22
purposes of, 81
tally sticks and, 79–84
objective, torsion of with subjective, 146, 171
"objective" leftism, 19
objective metaphysics, 88–90
objectivism, managed objectivism, 172
object-oriented philosophy, 164n12
O'Neil, Cathy, 29
ontology, naturality and, 201–6
opacity, 62–6
operaismo (workerism), 25
overfitting, 49, 50
overgeneralization, 49, 50

P
PageRank algorithm (Google), 78, 183–4
Panagia, Davide, 57n38
Papert, Seymour, 54
patents, 190n20
patriarchy
as constructed atop abstractions that separate concepts from context, 179
machine learning systems as working to solidify, 58
Pearson, Egon, 123–30, 136, 137, 138, 139, 144, 156, 164, 169, 172, 175, 189, 222
Pearson, Karl, 117, 160n3
Pennachin, Cassio, 43n11
percentage belief, 159–62
perceptron, 54
Perceptrons: An Introduction to Computational Geometry (Minsky and Papert), 54
Petabyte Age, 56
p-hacking, 134
Phillips, Lawrence, 144, 145
Pichai, Sundar, 202
Pierre-Dupuy, Jean, 30n55
Pitts, Walter, 53–4
Pius IX (pope), 71
platform capitalism, 26, 192–3
Plato, 164n12, 178
Poincaré, Henri, 103, 105
point estimates, 159n2
political economy

crises in statistics and capitalism as deriving from, 33
debates about nature of, 23–7
failure to account for, 220
growth in importance of to science, 133
knowledge as depending on, 222
statistics as functioning within and through, 175
political subject, question of viability of a unified political subject who could foment global revolutionary change, 16
Popper, Karl, 123, 124, 136
population genetics, 121
populations, statistical manipulation of, 29
posterior distributions, 159n2
Postone, Moishe, 13, 18, 27, 28, 76, 89n19, 208n54, 209
Price, Richard, 162–3, 164
primitive accumulation, 190, 191
privatization
of economy, 200
of general knowledge, 211, 218
neoliberalism's move toward, 3–4
probability
Bayesian analysis as admitting to theistic understanding of, 162
Dutch book argument as offering economic ground for mathematical laws of, 164
economic dynamics of, 105
frequentist description of, 104–5
inverse probability, 150
and management of contemporary capitalism, 39
as not directly assigned to hypothesis, 150
of precipitation (PoP), 102
requirements of to become objective, 146
understanding of, 101–4
ProPublica, 44
prostate-specific antigen (PSA) blood tests, USPSTF recommendation about, 127–8
Puar, Jasbir, 181
p-value (probability value), 117–20, 124, 138

R
race, UNESCO's statement on nature of, 121–2
racism
as allowing production of profit, 93
capital accumulation's constant drive to profit from, 220
centrality of to development and deployment of biopolitical control, 29
of COMPAS, 44
constant drive to profit from, 220
of Fisher, 121–3
as form of concrete domination, 92
machine learning systems as working to solidify, 58
as mobile and multifaceted system that defines varied axes of exchange, oppression, and injustice, 215
as not reducible to capitalism, 93
objectified forces as translating of into abstract demands of capital, 210
as presented in objective form, 174
racist housing policies, 204
as supporting injustice and exploitation, 75

radical political theory, crisis of, 15–16
radical politics
  fragmentation of, 18
  math, statistics, and science as necessary components of, 22
rationality, crisis of, 4
real abstraction, 11, 12, 13, 31, 75, 94, 139, 159, 223
redlining, 14, 204n48
reference classes, 40, 145–6
reference class problem, 40, 50
regulation, of algorithms, 60, 61
reification, theory of, 80n4, 82, 83n11
replication crises, 134, 160, 200, 219
re-simulation, methods of, 7
revolutionary mathematics
  aim of, 34, 224
  as attempting to create new objectifications, 214, 216
  as calling for reconsideration of relationships between alienation and liberation, 212–16
  need for, 33
  ultimate task of, 174–5, 176
  the work of, 33, 139–40
revolutionary object, theory of, 83n9
revolutionary political coalition, no unifying force as currently existing to hold it together, 18
revolutionary subject
  according to Marx, 15
  as beset by algorithmically fragmented reality and intensely managed digital control, 16
  as under siege and in doubt, 18
risk, according to Knight, 39
Robinson, Cedric, 91
Rosenblatt, Frank, 53–4

S
Sadowski, Jathan, 25n46
Saunderson, Nicholas, 143n1
Savage, Leonard, 103, 130, 146–7, 158, 169–70, 171, 189, 213, 220, 221, 222
Schelling, F.W.J., 49
Schrödinger, Erwin, 186–7
science
  as increasingly involving huge amounts of data, 2
  key question in, 220
scientific inference, 9, 114n23, 135, 160
scientific production, split with capitalism, 174
search space, 80
Serres, Michel, 133
Sevignani, Sebastian, 24n43, 209n59
sexism
  capital accumulation's constant drive to profit from, 220
  as form of concrete domination, 92
  as mobile and multifaceted system that defines varied axes of exchange, oppression, and injustice, 215
  as not reducible to capitalism, 93
  objectified forces as translating of into abstract demands of capital, 210
  as presented in objective form, 174
  as supporting injustice and exploitation, 75

Shiva, Vandana, 179
Smith, Adam, 190
Smith, George Davey, 118n7
Smythe, Dallas, 24n43, 88n16
socialism, as suffering from calculation problem, 195–6
social problems, 19, 60. *See also specific problems*
Socrates, 164n12
Sohn-Rethel, Alfred, 3, 11, 75, 139, 171, 179n3, 221
split tally sticks, 79–84
Srnicek, Nick, 25n46, 192
Standard 754 (IEEE), 68n14
statistical analyses
  ability of to evaluate diverse types of data, 119
  Bayesian approach as fundamentally changing stakes of, 160
  as facing possibility that what was observed was merely result of random chance, 109
statistical inference, 34, 108, 114n23, 120, 129, 148, 150, 199
Statistical Methods for Research Workers (Fisher), 115
statistical modeling, 41–3
statistical separability, theory of, 54
statistics
  ability of to revolutionize production, 10
  crisis of, 4, 33
  as effectively laundering nonobjective forms of violence and bias, 174
  epistemic weightiness of, 96
  Fisher-Neyman-Pearson model as, 131
  foundation of, 13, 171
  frequentist statistics. *See frequentist statistics/frequentism*
  goal of, 219
  hybridization of models of, 130–5, 137, 139
  as immaterial thing that can do everything, 31
  as increasingly governed by, 7
  and management of contemporary capitalism, 39
  metaphysical foundations of, 122
  as more efficient at creating new realities than in representing a world, 7
  as nothing short of magic, 175
  as objectifying, 31–2, 33
  as operating on two levels, 8–9
  as providing objectifying force of algorithmic knowledge, 88
  as providing tools for distinguishing signal from background noise, 132
  relation of to capitalism, 12–13
  role of in mediating society and economics "behind our backs," 20
  use of to create actionable information for human and computer consumption, 2
  as victim of its own success, 119
Sterne, Jonathan, 118n7
Stigler, Stephen, 143n1
stock market, 87
subjective, torsion of with objective, 146, 171
subjective belief, 94, 144, 148, 158, 163, 164, 166, 221
"subjective" leftism, 19
subjectivism, 146–8, 155, 160

supervised learning, 51, 52
surveillance capitalism, 26
systemic biases, algorithms and, 60

T

tally sticks, and objectification, 79–84
Target, use of data to predict customer behaviors, 44
targeted advertising, 10, 15, 28
Taylorist revolution/Taylorism, 10, 156
technology
    as being turned against humanity's very survival, 2
    negative consequences of, 3
    objectification as lying at heart of, 22
Teichmüller theory, 65
temporary correlation, 55
theoretical mathematics, and problem of opacity, 62–5
Theranos, 195
Timcke, Scott, 7n13
Tiqqun, 17
transparency, 61
transphobia, as mobile and multifaceted system that defines varied axes of exchange, oppression, and injustice, 215
Troncoso, Stacco, 95, 96
Trotsky, Leon, 141
Trump, Donald, 94
trust, as foundation of economic system, 4–5
truth
    crisis of, 4
    science of, 139
Turing, Alan, 78
Turing machine, 78
Tymoczko, Thomas, 63–4, 66, 70
type I errors, 107, 111, 126, 127, 151, 152
type II errors, 111, 124, 126

U

Uber
    "Greyball" program, 5–6
    success of, 192
    as taxing exchanges, 193
uncertainty, according to Knight, 39
UNESCO, statement on nature of race, 121–2
United States Preventive Services Task Force (USPSTF), recommendation about PSA testing, 127–8
universal form, 179
universal inference, 11
unsupervised learning, 51, 52
US Environmental Protection Agency, on Volkswagen's defeat device, 4–5

V

value, as deriving directly from automatic computation, 223
value extraction, nature of, 25
Venn, John, 40
Vernant, Jean-Pierre, 179n3
Virno, Paulo, 193, 198, 219, 220–1
Volkswagen, defeat device, 4–5, 6, 60
von Mises, Ludwig, 195–6, 197

W

Walmart
    hurricane data used by, 37–8, 41
    use of data to predict customer behaviors, 44
Wang, Jackie, 91–2
Weheliye, Alexander, 29
white supremacy
    as constructed atop abstractions that separate concepts from context, 179
    objectification of into statistical matrix of one's property value, 204
    reproduction and incentivization of, 93
"Why Most Published Research Findings Are False" (Ioannidis), 135, 139
Wynter, Sylvia, 203

X

xenophobia, as mobile and multifaceted system that defines varied axes of exchange, oppression, and injustice, 215

Y

YouTube
    use of videos as training data for Google's image recognition system, 186, 187
    video recommendation algorithms, 92–3

Z

Žižek, Slavoj, 23n42, 60n2